Discrete Wavelet Transform: A Game Changer for Radio Frequency Interference Mitigation

Ravi

Copyright © 2024 by Ravi

Table of Contents

Chapter 1

Introduction

1.1 Background to the Study

Radio-frequency interference (RFI) is a electromagnetic disturbance that occurs in the radio-frequency spectrum that causes the degradation of signal quality in a system. These disturbances may be caused by both man-made and natural sources and effect various electromagnetic systems such as radars, mobile phones as well as radio-telescopes. For this reason an investigation into RFI mitigation using wavelets was done.

A wavelet is a signal defined for some brief period of time that contains oscillatory characteristics. Generally, wavelets are intentionally designed to posses particular qualities relevant to a certain signal processing application.

Currently, the body of research relating to wavelets has been fairly underutilized in interference removal, where the preferred techniques being used are adaptive filtering and more classical frequency domain based techniques. However, with the advances in the field of wavelets, beginning with Ingrid Daubechies' research into the topics of maximally flat wavelet filters to the parametrization of wavelet filter coefficients using trigonometric functions have made the field of wavelets unprecedentedly suitable for the task of interference suppression.

Therefore, this research project attempts to prove that the suppression of interference using the Discrete Wavelet Transform (DWT) is highly effective. The result of this research is an adaptive wavelet-basis, interference suppression system that has been applied to radio astronomy imaging.

1.2   Objectives of the Study

This research project attempts to prove the viability and effectiveness of wavelet based interference removal techniques. In order to do so the following objectives were set:

- Review existing literature and determine the most feasible wavelet based interference mitigation technique.

- Using the optimal wavelet-based approach, design suitable wavelets for general interference excision.

- Prove the validity of the designed wavelets using performance heuristics.

- Integrate the designed wavelets into an algorithmic library-based interference suppression system.

- Apply the wavelet based suppression scheme to radio astronomy data sets collected by the KAT-7 [1] radio-telescope array and demonstrate its performance using imaging.

- Measure the performance of the RFI excision based on the imaged data and draw conclusions based on the effectiveness of the suppression system.

1.3   Motivation for the Study

This research project may be motivated from a macroscopic view of the effective usage of wavelets within field of interference mitigation techniques, as well as the microscopic view of attempting to remove interference for remote sensing and radio astronomy applications.

Macroscopically, this research can be motivated by wavelets being a severely underutilized tool in the fields of signal processing and remote sensing therefore any additional research to this field may be beneficial to both the wavelet and remote sensing communities. Furthermore, microscopically this research is motivated by the need to prevent interference from corrupting the research in radio astronomy.

1.4 Scope and Limitations

As this research project aims to prove the validity and effectiveness of wavelets in the context of general interference suppression it necessary to limit the scope of the project. Therefore, this project is focused on the design and use of one dimensional discrete wavelet methods that ensure the orthogonality of basis functions. Furthermore, the implementation of the DWT is limited to the use of perfect reconstruction (PR) filter-banks [2] without using the lifting scheme [3].

Finally the validation of the designed algorithm is restricted to the observation of the Messier 83 (M83) [4] galaxy collected by the KAT-7 radio-telescope.

1.5 Plan and Development

This report begins with an overview of existing wavelet theory and the mathematics used through out this research. The next chapter documents literature pertaining to interference suppression schemes and chapter four documents the design procedure of a smooth orthonormal library of wavelets made for interference excision.

Chapter five then addresses the library's integration into a larger interference suppression scheme that is applied to test interference data for validation. The following chapter documents the application of the interference suppression system to KAT-7 observations of the M83 galaxy. Chapter seven documents the results from the radio astronomy imaging based application and chapters eight and nine draw conclusions and recommendations based on these results.

The plan and development of this research project may be seen in v-diagram in Appendix A.1.

Chapter 2

Wavelet Theory

This chapter provides an overview of the theory used for wavelet based interference suppression in this research project. Its purpose is to give the reader intuition regarding the properties of both the discrete and continuous wavelet transforms, so that the exact details may be abstracted in later chapters.

2.1 Background to Signal Processing

Signal processing is the field related to the analysis and modification of signals. Commonly signals appear in the time domain, and offer some information about the physical environment from which they were observed or recorded.

The Fourier transform is a well known method used to transform an observed time domain signal to the frequency domain in order to obtain more information about its composition. This is done by considering the similarity of the observed signal with a series of sinusoids at different frequencies, such that the resulting transformation contains information only about the frequency spectrum of the transformed signal and none about its time localization. The mathematical formulation of the Fourier transform can be expressed by

$$X(\omega) = \int_{-\infty}^{\infty} x(t)e^{-j\omega t}dt \tag{2.1}$$

where $X(\omega)$ is the frequency domain representation of the time domain signal $x(t)$ and $\omega = 2\,\pi f$. In addition to this, multiplication in the frequency domain can be expressed by convolution in the time domain.

In order to give signal analysts more information about the time/frequency localization of particular features an observed signal, the short-time Fourier Transform (STFT) is often used. This is achieved, by partitioning the observed signal into windows and finding the frequency content of each window, thereby localizing frequency information to a particular time interval.

More recently, the wavelet transform has been used for the purpose of time/frequency analysis. The wavelet transform can be segmented into two different classes, the Continuous Wavelet Transform (CWT) and the Discrete Wavelet Transform (DWT). The particular differentiating factors of each transform will be expressed in later explanations.

The CWT is comparable to the STFT in that it finds the similarity of an observed signal with an analysis function at a particular scale or frequency. In the case of the wavelet transform, wavelets are used as the analysis functions which are functions that are defined for some period in time which have oscillatory characteristics. An example of three different types of wavelets can seen in Figure 2.1.

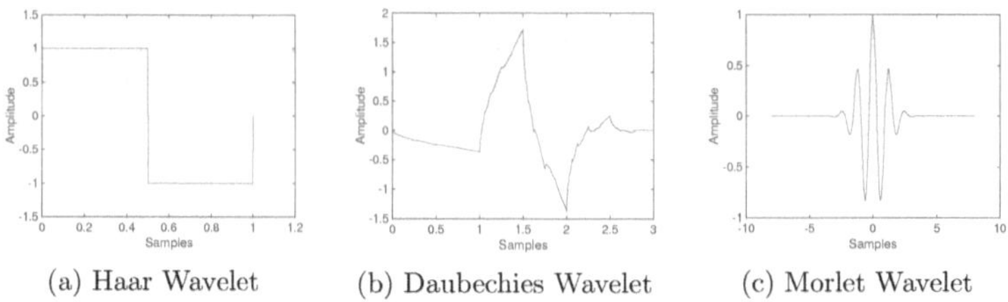

(a) Haar Wavelet   (b) Daubechies Wavelet   (c) Morlet Wavelet

Figure 2.1: Figure Documenting Three Different Types of Wavelets

The key difference between Fourier and continuous wavelet based techniques is that wavelets have good time and frequency localization whereas in Fourier based techniques, the analysis functions are sinusoids that have good frequency resolution but poor time resolution. On the contrary, the DWT uses nested filter-banks to perform multi-resolution analysis, which as per the name, is the analysis of a particular signal at various frequency and time resolutions.

5

This chapter explains the underlying mathematics and theory of wavelet transforms as well as the tools used for interference excision.

2.2 The Continuous Wavelet Transform

Although this research project focuses on the application of discrete wavelets to interference mitigation techniques, the continuous wavelet domain offers great insight into the mechanics of the wavelet transform and wavelet theory. For this reason, the CWT is covered for purposes of giving the reader intuition about wavelets so that the later application based sections may speak more abstractly about the details of the decompositions.

Simply put, the continuous wavelet transform convolves an analysis function at varying scales with a particular signal to be analyzed. This operation can be expressed mathematically by

$$X_\omega(s, \tau) = \frac{1}{\sqrt{s}} \int_{-\infty}^{\infty} x(t)\psi^* \left(\frac{t - \tau}{s}\right) dt. \tag{2.2}$$

Where the two variables s and τ are the respective scaling and translation factors of the wavelet transform. Changes in the scale of a wavelet, s, cause the analysis function to be contracted or stretched along the time axis thereby varying the frequency content of the analysis function. This process is called *dilating* or *scaling* as it allows for analysis of a signal at varying resolutions.

By altering the time shift, τ, of the wavelet allows for the analysis of a signal at different instances in time, thereby *sliding* the analysis function along the time axis. In addition to this it must be noted that the $1/\sqrt{s}$ term ensures that the wavelet analysis conforms to the normalization condition at every scale.

This being said, in order to properly define the continuous wavelet transform, it is necessary to explain the requirements of a wavelet. In more formal terms a wavelet, $\psi(t)$, is a function in $\mathbb{L}^2(\mathbb{R})$ space, that satisfies the following conditions [5]:

1. It is absolutely square and integrable:

$$\int_{-\infty}^{\infty} |\psi(t)|^2 dt \leq \infty$$

2. It has a zero mean:

$$\int_{-\infty}^{\infty} (t)dt = 0$$

3. It has a norm of 1:

$$||\psi(t)|| = 1$$

where the first condition is the definition of $\mathbb{L}^2$ space. Lebesgue-2 space is the space of all functions that have a defined integral of the square of the modulus of the function. Such that, $\mathbb{L}^2(\mathbb{R})$ is the subset of $\mathbb{L}^2$ that is contained by real numbers.

The second condition relates to the fact that wavelets are required to have compact support. The final condition ensures that the wavelet does not distort the signal being analyzed and is referred to as the normalization condition.

In addition to this, some literature such as that by Merry [6], define frequency domain based constraints such that the wavelet has no DC offset and that the norm of the Fourier transform of the wavelet must have a defined integral between 0 and ∞. This ensures that the wavelet has band-pass characteristics in the frequency domain, which is a very useful property for various applications such as non destructive filtering and analysis of signals.

As shown in Figure 2.1, there exist a variety of different wavelets, each with different shapes in the time domain. One of the effects of this is that the center frequency of each wavelet varies significantly, thereby making it hard to formalize the relationship between scale and frequency for each dilation of a wavelet in the CWT. The center frequency of a wavelet is most easily understood as the most prominent frequency domain component of the lowest scale of a wavelet function.

In the case of the Morlet wavelet shown in Figure 2.1c, the frequency at each scale may be represented as [6]

$$f_c = \frac{\eta}{s} \tag{2.3}$$

where η is the center frequency of the wavelet and s is the scale variable. In much of wavelet literature, this relationship is used to give insight into the scale phenomenon associated with wavelets. The relationship is normally approximated as $f_c \propto 1/s$.

An example of the the way in which the change in scale and time effects the frequency and shape of the wavelets is shown in Figure 2.2. It can be seen that at scale s the wavelet is squashed in time and that there is good time localization but poor frequency localization. Whereas at scale s_0, the wavelet is stretched in time thereby leading to poor time resolution but high frequency resolution.

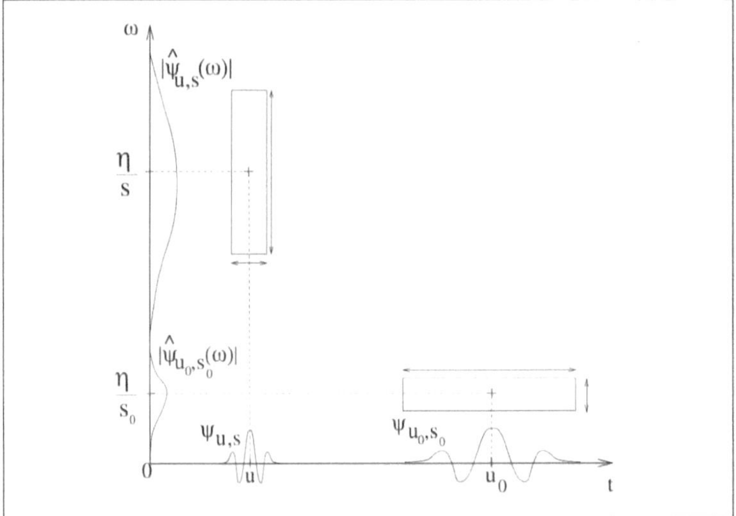

Figure 2.2: Figure Showing the Effect of Dilating and Shifting Wavelets in the Time/Frequency Plane [7]

It must be noted that in most implementations of the CWT in common software packages such as MATLAB [8], the output scale/time plot is reversed along the scale axis. Where high scales (low frequencies) are plotted low on the scale axis whereas lows scales (high frequencies) are plotted in the higher positions on the scale axis.

In order to effectively transition into the theory surrounding the DWT, it is necessary to consider the relationship between the CWT and the Discrete Time Wavelet Transform (DTWT). Although their similarities are useful to describe the relationship between the two domains, the derivation of the DWT in later sections is somewhat non-linear.

This being said, if the scale parameter s from Equation (2.2) is substituted by a quantized variable such that $s = 2^{-j}$ and the continuous time parameter t parameter is replaced by a sampled time parameter k, then the result of this is [9]

$$X_\omega(j, k) = 2^{j/2} \sum_n x(n)\psi^*(2^j n - k). \tag{2.4}$$

It must be noted that in order for the DTWT and CWT to comparable the transform must be critically sampled, meaning that the quantized variable must be sampled inversely to 2 time the maximum frequency of $x(n)$. This ensures that that all information present in the CWT decomposition is the same as that of the DTWT [10].

This section may be summarized by considering the differences between the DWT and CWT. The most critical difference between them is that the CWT often results in a redundant projection of a signal, where the wavelets at each scale are non-orthogonal to each other. Corollary, the DWT ensures orthogonality of the analysis functions so that the expansions of a signal at each scale share no similarities.

The effect of this is that the CWT is well suited to analytical applications for its capacity to continuously span the time-frequency (scale) domain where the the inverse transform is not required. With applications to fields such as geophysics, acoustics and system identification [11].

Whereas applications of the DWT are those which require no redundancy of the expansion of signals, with uses such as data compression and noise reduction. The precise details of these applications will be documented in later sections.

2.3 Multiresolution Analysis

The theory surrounding discrete wavelets is highly independent to its continuous analogue. The easiest approach to describe the topic is by considering the functional spaces that a wavelet at each scale contributes to. This functional segmentation is referred to as Multiresolution Analysis (MRA). In order to gain an understanding of this, it is necessary to consider the field of functional analysis so to understand how basis make up various functional spaces.

Much of the theory in this chapter surrounding multiresolution analysis appears in the work by Burrus [12].

2.3.1 Signal Spaces

In order to accurately describe the phenomena of multiresolution analysis, it is necessary to define terminology from functional analysis. A signal or function, $f(t)$, can often be better analyzed if expressed as a linear decomposition by

$$f(t) = \sum_k a_k \varphi_k(t) \tag{2.5}$$

where k is an integer index, a_k is real-valued expansion coefficients and $\varphi_k(t)$ is a set of real-valued analysis functions of t. If the expansion is unique, then the set a_k is called a basis for the class of functions that can be expressed. If the basis is orthonormal, then we can say that

$$\langle \varphi_k(t), \varphi_j(t) \rangle = \int \varphi_k(t)\varphi_j(t)dt = 0 \qquad \text{where } k \neq j \tag{2.6}$$

this means that it is possible to calculate the expansion coefficients using the inner product, such that

$$a_k = \langle f(t), \varphi_k(t) \rangle = \int f(t)\psi_k(t)dt. \tag{2.7}$$

Where the norm or length of a vector is given by

$$\|f(t)\| = \sqrt{|\langle f, f \rangle|} \tag{2.8}$$

which is a generalization of the geometric operations in Euclidean space. From this definition, it is possible to say that two vectors or signals are orthogonal if their product is zero and their norms are non-zero.

2.3.2 Scaling Functions

In order to effectively make use of MRA, the scaling function, $\varphi(t)$, is first considered. To formalize notation k is used to define integer translates of $\varphi(t)$ such that

$$\varphi_k(t) = \varphi(t - k), \quad k \in \mathbb{Z} \quad \text{and} \quad \varphi \in \mathbb{L}^2 \tag{2.9}$$

which means that the subspace of $\mathbb{L}^2(\mathbb{R})$ spanned by $\varphi_k(t)$ is

$$\mathcal{V}_0 = \overline{\operatorname{Span}_k \varphi_k(t)}, \quad k \in \mathbb{Z} \tag{2.10}$$

where the over-line denotes the closure of this space. The implication of this is that

$$f(t) = \sum_k a_k \varphi_k(t), \quad \text{for any} \quad f(t) \in \mathcal{V}_0 \tag{2.11}$$

This result means that one can change the subspace that $\varphi(t)$ spans by changing the position on the time axis that scaling function occupies. Furthermore, this can be extended using the scaling phenomena described in Equation (2.2) as denoted by j, which can be expressed as

$$\varphi_{j,k}(t) = 2^{j/2} \varphi(2^j t - k) \tag{2.12}$$

such that $\varphi_{j,k}(t)$'s span is

$$\mathcal{V}_j = \overline{\operatorname{Span}_k \varphi_k(2^j t)} = \overline{\operatorname{Span}_k \varphi_{j,k}(t)}, \quad k \in \mathbb{Z} \tag{2.13}$$

which implies that if $f(t)$ is a function of $\mathcal{V}_j$ then it can be expressed by

$$f(t) = \sum_k a_k \varphi_{j,k}(t). \tag{2.14}$$

In order to generalize the results shown, it is necessary to formulate the requirements of the MRA by nesting the resulting subspaces. This can be shown as

$$\cdots \subset \mathcal{V}_{-2} \subset \mathcal{V}_{-1} \subset \mathcal{V}_0 \subset \mathcal{V}_1 \subset \mathcal{V}_2 \cdots \subset \mathbb{L}^2 \tag{2.15}$$

such that

$$\mathcal{V}_j \subset \mathcal{V}_{j+1}, \forall j \in \mathbb{Z} \tag{2.16}$$

so that the subspace $\mathcal{V}_{-\infty}$ only contains 0 or the null subspace and $\mathcal{V}_\infty$ contains all functions in $\mathbb{L}^2$. This means that the space that contains high resolution functions will also contain low resolutions functions, but the space that only contains low resolution functions will not contain high resolution

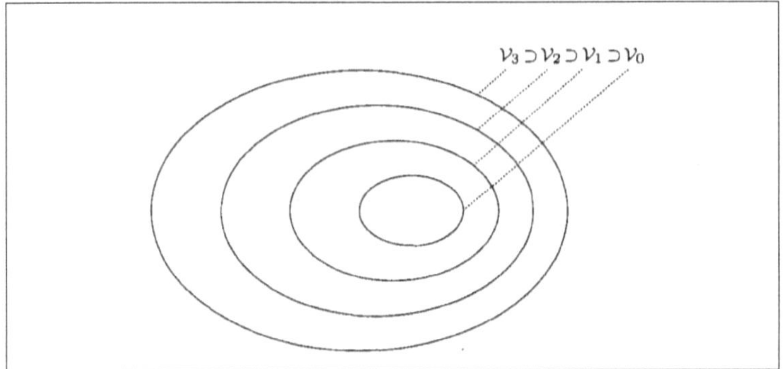

Figure 2.3: Nested Functional Spaces Spanned by Scaling Functions [13]

functions. This means that some function, $f(t)$, is contained by $\mathcal{V}_j$ whereas $f(2t)$ is contained in the subset $\mathcal{V}_{j+1}$. An illustration of this concept may be seen in Figure 2.3.

The implications of this are that $\varphi(t)$ can be expressed in terms of a weighted sum of shifted $\varphi(2t)$ by

$$\varphi(t) = \sum_n h_0(n)\sqrt{2}\varphi(2t - n) \tag{2.17}$$

where $h_0(n)$ is some coefficients that will be shown later to correspond to a low pass filter and $\sqrt{2}$ is used to maintain unit energy in each scaling function. This equation is fundamental to our construction of discrete wavelets and is known as the dilation equation or the refinement equation [13].

2.3.3 Wavelet Functions

In most cases, the important features of a signal are usually contained in the higher resolutions spaces spanned by the scaling functions, $\varphi_{j,k}$, when j is very large. However these features may be better described by not, $\varphi_{j,k}$, but rather by using a different set of functions that span the differences of the spaces spanned by the scaling functions. These functions are called the wavelet functions and are denoted by, $\psi_{j,k}$, where j and k denote the dilations and shifts in time respectively.

An important feature of wavelet functions is that they are orthogonal to scaling functions, this relationship is expressed by

$$\langle \varphi_{j,k}, \psi_{j,l} \rangle = \int \varphi_{j,k} \psi_{j,l} dt = 0, \quad j,k,l \in \mathbb{Z}. \tag{2.18}$$

There are two advantages of preserving orthogonality between wavelet and scaling functions. The first is that the calculation of expansion coefficients is easier through the use of a filter bank approach (the details of which will be described in later sections). The second reason for preserving orthogonality is so the DWT becomes a unitary transform. This means that the energy in the original signal is the same in that of the coefficients of the transformed signal as per Parseval's Theorem shown in Equation (2.29).

As already mentioned, the space spanned by, $\varphi_{j,k}(t)$, is $\mathcal{V}_j$, the space which is orthogonal to it is spanned by, $\psi_{j,k}(t)$, and is referred to as $\mathcal{W}_j$. This means that all functions in $\mathcal{V}_j$ are orthogonal to all functions in $\mathcal{W}_j$. From this it is possible to formulate the wavelet spanned subspace $\mathcal{W}_0$ such that

$$\mathcal{V}_1 = \mathcal{V}_0 \oplus \mathcal{W}_0 \tag{2.19}$$

which extends to

$$\mathcal{V}_2 = \mathcal{V}_0 \oplus \mathcal{W}_0 \oplus \mathcal{W}_1 \tag{2.20}$$

which can be generalized by

$$\mathbb{L}^2 = \mathcal{V}_0 \oplus \mathcal{W}_0 \oplus \mathcal{W}_1 \oplus \mathcal{W}_2 \oplus \ldots \tag{2.21}$$

It can be noted that the scale of the initial subspace is arbitrary, as higher order subspaces contain lower order subspaces. This implies that

$$\begin{aligned}
\mathbb{L}^2 &= \mathcal{V}_{10} \oplus \mathcal{W}_{10} \oplus \mathcal{W}_{11} \oplus \ldots \\
&= \mathcal{V}_{-15} \oplus \mathcal{W}_{-15} \oplus \mathcal{W}_{-14} \oplus \ldots \\
&= \cdots \oplus \mathcal{W}_{-2} \oplus \mathcal{W}_{-1} \oplus \mathcal{W}_0 \oplus \mathcal{W}_1 \oplus \mathcal{W}_2 \ldots
\end{aligned} \tag{2.22}$$

An illustration of this concept can be seen in Figure 2.4.

Since the wavelets reside in the space spanned by the next narrower scaling function, $\mathcal{W}_0 \subset \mathcal{V}_1$, they can be represented by sum of weighted and shifted scaling functions, $\varphi(2t)$ as defined by

$$(t) = \sum_n h_1(n)\sqrt{2}\varphi(2t - n), \quad n \in \mathbb{Z}. \tag{2.23}$$

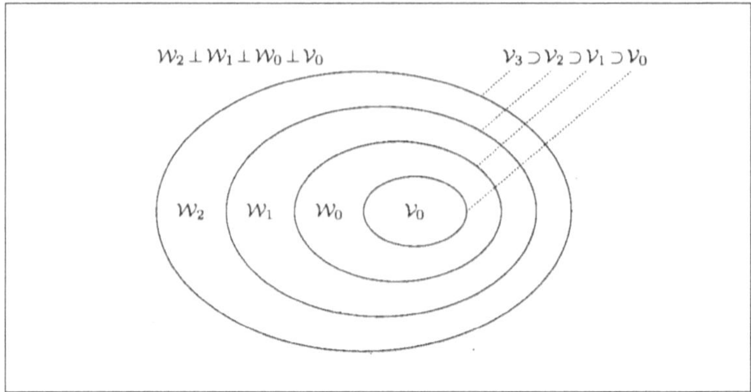

Figure 2.4: Nested Functional Spaces Spanned by Scaling Functions and Wavelet Functions [13]

where $h_1(n)$ is some coefficients that will be shown later to correspond to a high pass filter. Furthermore, from the result of Equations (2.23) and (2.22) we can say that

$$\psi_{j,k}(t) = 2^{j/2}\psi(2^{j/2}t - k) \tag{2.24}$$

Finally, it can be noted that, the wavelet function leads to a more detailed analysis of a function than the scaling function. For this reason, we refer to the scaling function as a low pass filter, as it smooths details in a signal just as a moving average filter does. Whereas the wavelet functions result in finer details of the signal being analyzed and is therefore considered a high pass filter [13].

2.4 The Discrete Wavelet Transform

In order to elaborate on the idea that the DWT has stringent conditions on orthogonality and non-redundant padding in the time-scale domain it is necessary to mathematically formulate this conditions. This sections considers the mathematical structure of these conditions as well as the filter bank based implementation of the DWT.

2.4.1 DWT Definition

Considering the expression given in Equation (2.22), it can be said that

$$\mathbb{L}^2 = \mathcal{V}_{j0} \oplus \mathcal{W}_{j0} \oplus \mathcal{W}_{j0+1} \oplus \dots \qquad (2.25)$$

then through the use of Equations (2.17) and (2.23) it is possible to generalize the expansion of some signal, $g(t)$, by using wavelets and scaling functions such that

$$g(t) = \sum_{k} c_{j_0}(k)\varphi_{j_0,k}(t) + \sum_{k}\sum_{j=j_0}^{\infty} d_j(k)\psi_{j,k}(t). \qquad (2.26)$$

Where the value of j_0 sets the coarsest scale whose space is spanned by the scaling function $\varphi_{j_0,k}(t)$. The rest of the $\mathbb{L}^2(\mathbb{R})$ is spanned by the wavelet functions $\psi_{j,k}(t)$ which provide the fine details of the function, $g(t)$. The coefficients of the DWT, $c_{j_0}(k)$ and $d_j(k)$ describe the similarity to the signal under analysis to some basis. If the wavelet system is orthogonal then we can calculate the coefficients by

$$c_j(k) = \langle g(t), \varphi_{j,k}(t) \rangle = \int g(t)\varphi_{j,k}dt \qquad (2.27)$$

which are known as the approximation coefficients and

$$d_j(k) = \langle g(t), \psi_{j,k}(t) \rangle = \int g(t)\psi_{j,k}dt \qquad (2.28)$$

are known as the detail coefficients [13].

2.4.2 Parseval's Theorem

As already mentioned, if the scaling and wavelet functions form an orthonormal basis then Parseval's Theorem relates the energy contained in each component of the DWT coefficients to the energy of the original signal. This is shown by

$$\int |g(t)|^2 dt = \sum_{l=-\infty}^{\infty} |c(l)|^2 + \sum_{j=0}^{\infty}\sum_{k=-\infty}^{\infty} |d_j(k)|^2 \qquad (2.29)$$

where the energy in the wavelet domain partitioned by scale, j, and time k. This is an important condition because it proves that the energy contained in the signal under analysis is equivalent to the energy of the detail and approximation coefficients of the DWT. This in turn enables the use of the DWT in applications such as compression and interference suppression.

2.4.3 Analysis using Filter Banks

In most discrete applications, it is convenient to not have to deal directly with the wavelet and scaling functions. Thereby only making use of the the approximation and detail coefficients, $c_j(k)$ and $d_j(k)$ and the coefficients governing them $h_0(n)$ and $h_1(n)$. It will be seen that this can be achieved through the use of filter banks and sub-samplers.

In order to work directly with the DWT coefficients, the relationship between the filter coefficients and the detail coefficients is derived by

$$\varphi(t) = \sum_n h_0(n)\sqrt{2}\varphi(2t-n) \tag{2.30}$$

assuming a unique solution exists, we dilate $\varphi(t)$ in time by k and scale it by 2^j such that

$$\varphi(2^j t - k) = \sum_n h_0(n)\sqrt{2}\varphi(2(2^j t - k) - n) = \sum_n h_0(n)\sqrt{2}\varphi(2^{j+1}t - 2k - n). \tag{2.31}$$

Then if we set $m = 2k + n$ it can be seen that

$$\varphi(2^j t - k) = \sum_m h_0(m - 2k)\sqrt{2}\varphi(2^{j+1}t - m) \tag{2.32}$$

so that if we consider Equation (2.13) then it can be said that if

$$f(t) \in \mathcal{V}_{j+1} \tag{2.33}$$

then

$$f(t) = \sum_k c_{j+1}(k)2^{(j+1)/2}\varphi(2^{j+1}t - k). \tag{2.34}$$

Which means that $f(t)$ can be expressed at a scale of $j+1$, using only scaling functions. However, if we wanted to express $f(t)$ using scaling functions and wavelets it would be only necessary to express $f(t)$ at the scale of j. Therefore we can say that

$$f(t) = \sum_k c_j(k)2^{j/2}\varphi(2^j t - k) + \sum_k d_j(k)2^{j/2}\psi(2^j t - k) \tag{2.35}$$

where the $2^{j/2}$ term maintains unit energy of the basis function at various scales. Furthermore, it can be seen that Equations (2.35) and (2.34) are equivalent. If $\varphi_{j,k}(t)$ and $\psi_{j,k}(t)$ are orthonormal then the j level scaling coefficients can be found by

$$c_j(k) = \langle f(t), \varphi_{j,k}(t) \rangle = \int f(t)2^{j/2}\varphi(2^j t - k)dt \qquad (2.36)$$

then if we use Equation (2.30) and reorder the summation and integral then it can be seen that

$$c_j(k) = \sum_m h_0(m - 2k) \int f(t)2^{(j+1)/2}\varphi(2^{j+1}t - m)dt. \qquad (2.37)$$

From this result, we can say that

$$c_j(k) = \sum_m h_0(m - 2k)c_{j+1}(m) \qquad (2.38)$$

and the corresponding relationship for the wavelet coefficients is given by [13]

$$d_j(k) = \sum_m h_1(m - 2k)c_{j+1}(m) \qquad (2.39)$$

Filtering and Downsampling

In signal processing, filtering is the process of convolving a sequence of numbers with a particular signal in order to highlight certain features of a signal. In the field of multirate filters, relationships are exploited surrounding the information contained at particular indexes of signal. Two basic operations in multirate filters are the downsampler and the upsampler. The downsampler or decimator can be described by

$$y(n) = (\downarrow 2)x(n) = x(2n) \qquad (2.40)$$

where $x(n)$ is the input signal, $y(n)$ is the output signal and $\downarrow 2$ represents downsampling at a rate of 2. This means that every second sample of x(n) is discarded when put through a decimator [2].

The reason that filters and downsampling are necessary concepts for the DWT is that the discrete wavelet expansion can be implemented using these concepts. If we consider Equations (2.38) and (2.39) it can be seen that the scaling and wavelet coefficients at different levels can be obtained by convolving the expansion coefficients at scale j with the time reversed recursion coefficients $h_0(-n)$ and $h_1(-n)$ and then downsampling by 2 to give the expansion coefficients at the next level $j + 1$ [13].

These structures implement Mallat's Algorithm [7], and have been developed using ideas surrounding filter banks, quadrature mirror filters (QMF) and perfect reconstruction filter banks, which will be covered in later sections.

Fig. 2.5 shows a two stage, two band analysis bank, which can be used to implement a DWT with 2 levels of decomposition. Whereas Fig. 2.6 shows a three stage, two band DWT with 3 levels of decomposition. The functional spaces that each of the levels occupy can be seen in this figure. Where $\mathcal{V}_3$ can be decomposed into $\mathcal{W}_2$, $\mathcal{W}_1$, $\mathcal{W}_0$ and $\mathcal{V}_0$.

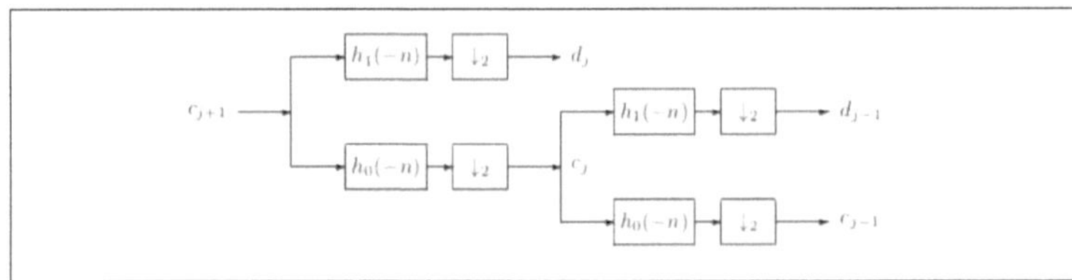

Figure 2.5: A Two Stage Two Band DWT Analysis Bank [13]

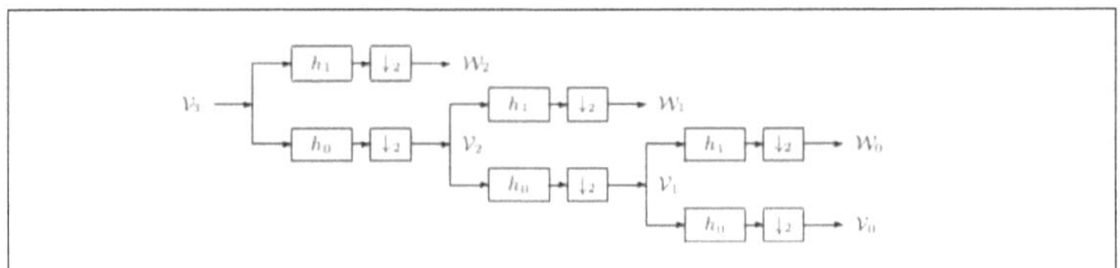

Figure 2.6: A Three Stage Two Band DWT Analysis Bank [13]

In order to gain an intuitive understanding of the processes taking place in Fig. 2.5 and 2.6 it is necessary to view the frequency domain based responses of each of the stages.

The first stage of banks divides the spectrum of c_{j+1} into a low pass and a high pass band resulting in the scaling and wavelet coefficients at a lower scale c_j and d_j. The next stage divides the resulting low pass spectrum into another low pass and a band pass spectra. This logarithmic set of bandwidths can be seen in Fig. 2.7. It can be noted that this description of partitioning of frequency bands obeys the rule defined in Equation (2.29).

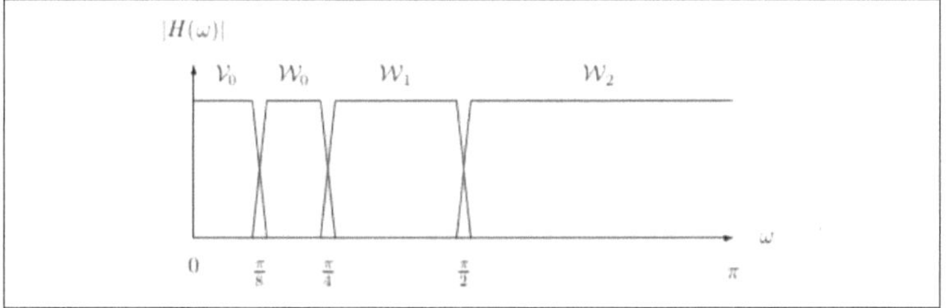

Figure 2.7: Frequency bands for the Analysis Tree [13]

It can be noted that for any band-limited signal there will be an upper scale of $j = J$ above which the detail coefficients, $d_j(k)$ are negligibly small [14]. By starting with a high resolution description of a signal in terms of the scaling coefficients, c_J, the analysis tree calculates the DWT down to $j = j_0$ such that there are $J - j_0$ stages [13].

2.4.4 Synthesis using Filter Banks

In the same manner that the analysis stage represented a signal with smaller scales, the synthesis stage takes the smaller scales that are used to represent a signal and recombines them to form the original signal. This process is derived by considering the $j + 1$ scaling space which is given by

$$f(t) = \sum_k c_{j+1}(k)2^{(j+1)/2}\varphi(2^{j+1}t - k) \tag{2.41}$$

which may be represented in terms of the next scale as

$$f(t) = \sum_k c_j(k)2^{(j)/2}\varphi(2^j t - k) + \sum_k d_j(k)2^{(j)/2}\psi(2^j t - k). \tag{2.42}$$

Then by substituting Equation (2.30) and (2.23) into (2.42) , it can be seen that

$$\begin{aligned} f(t) = &\sum_k c_j(k) \sum_n h_0(n)2^{(j+1)/2}\varphi(2^{j+1}t - 2k - n) \\ &+ \sum_k d_j(k) \sum_n h_1(n)2^{(j+1)/2}\varphi(2^{j+1}t - 2k - n) \end{aligned} \tag{2.43}$$

then by equating Equations (2.41) and (2.43), setting $m = 2k - n$ and finding their inner products with $\varphi_{j+1,k'}(t)$, it can be seen that

$$\sum_k c_{j+1}(k) \int \varphi(2^{j+1}t - k)\varphi(2^{j+1}t - k)dt$$

$$= \sum_k c_j(k) \sum_n h_0(n) \int \varphi(2^{j+1}t - k)\varphi(2^{j+1}t - m)dt \qquad (2.44)$$

$$+ \sum_k d_j(k) \sum_n h_1(n) \int \varphi(2^{j+1}t - k)\varphi(2^{j+1}t - m)dt$$

which means that after some rearranging of Kronecker deltas and changing of indices, the synthesis equation can be given by [13]

$$c_{j+1}(k) = \sum_m c_j(m)h_0(k - 2m) + \sum_m d_j(m)h_1(k - 2m). \qquad (2.45)$$

It must be noted that the indices for synthesis equation given in (2.45) are the inverse of the indices of the analysis equations given in (2.38) and (2.39). From this result it is possible to infer that upsampling by 2 is required, as the analysis stages needed downsampling by 2.

The process of upsampling can be phrased mathematically by

$$y(n) = (\uparrow 2)x(n) \implies y(2n) = x(n) \quad \text{and} \quad y(2n + 1) = 0 \qquad (2.46)$$

if $x(n)$ is the input to the upsampler and $y(n)$ is the output. This means that the process of upsampling inserts zeros into every odd index of a signal. Therefore the synthesis of scaling coefficients can be achieved through the use of filter banks as shown in Fig. 2.8.

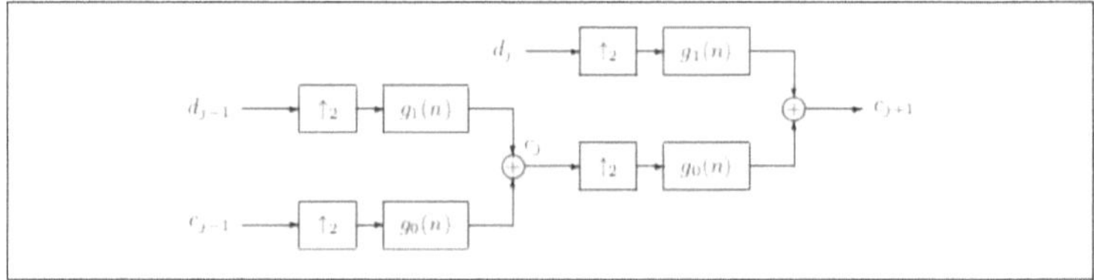

Figure 2.8: Two Stage, Two Band Synthesis Tree [13]

In conclusion of the discrete wavelet theory shown, it can be said that the DWT enables orthogonal decompositions and recomposition of signals. This property is the reason why wavelet have become so revered in the signal processing community, as this was the first tool to achieve this.

2.5 Properties of Filter Banks

In order to gain a better understanding of the conditions required for the design of scaling and wavelet coefficients relating to a particular DWT, it is necessary to investigate filter banks. The analysis done in this section will consider the frequency response of the filter banks, thereby allowing us to infer properties that relate the low and high pass filters to the DWT.

2.5.1 Subsampling in the Fourier Domain

Downsampling

If we consider the signal

$$y(n) = (\downarrow 2)x(n) \tag{2.47}$$

implies that

$$y(n) = x(2n) \tag{2.48}$$

then the Fourier Transform of $y(n)$ is

$$Y(\omega) = \sum y(n)e^{-jnw} = \sum x(2n)e^{-jnw}. \tag{2.49}$$

Now consider, $u(n)$, to be the same as, $x(n)$ except with its odd numbered components set to zero such that

$$u(n) = \begin{cases} x(n) & \text{if} \quad n \quad \text{is even} \\ 0 & \text{if} \quad n \quad \text{is odd} \end{cases} \tag{2.50}$$

which means that

$$(\downarrow 2)u(n) = (\downarrow 2)x(n). \tag{2.51}$$

Therefore the Fourier transform of $u(n)$ can be said to be

$$U(\omega) = \sum_{n \text{ even}} x(n)e^{-jnw} = \frac{1}{2}\sum_n x(n)e^{-jnw} + \frac{1}{2}\sum_n x(n)e^{-jn(w+\pi)} \tag{2.52}$$

this is because the addition of π to the frequency, reverses the signs of the odd-numbered components. The result of this is

$$U(\omega) = \frac{1}{2}[X(\omega) + X(\omega + \pi)] \tag{2.53}$$

which can be used to show that

$$Y(\omega) = U(\frac{\omega}{2}) = \frac{1}{2}[X(\frac{\omega}{2}) + X(\frac{\omega}{2} + \pi)]. \tag{2.54}$$

Upsampling

If we now consider

$$u(n) = (\uparrow 2)x(n) \quad \text{implies that} \quad \begin{cases} u(2n) & = x(k) \\ u(2n+1) & = 0 \end{cases} \tag{2.55}$$

it can be seen that all odd samples of $x(n)$ are zeros, and all even samples are the signal $x(n)$. This means that the odd terms contribute to no frequency components of $U(\omega)$. The implications of this are that

$$\begin{aligned} U(\omega) &= \sum_n u(n)e^{-jn\omega} = \sum_n u(2n)e^{-j2n\omega} \\ &= \sum_n x(n)e^{-j2n\omega} = U(2\omega). \end{aligned} \tag{2.56}$$

Such that when a signal is first upsampled and then downsampled, the following holds [2]

$$\mathcal{F}\{(\downarrow 2)(\uparrow 2)u(n)\} = U(\omega) = \frac{1}{2}[X(\omega) + X(\omega + \pi)]. \tag{2.57}$$

2.5.2 Subsampling in the Z-Domain

The z-transform is a complex frequency domain transform that operates on discrete-time signals. It is mathematically defined by

$$X(z) = \sum_{n=-\infty}^{\infty} x(n)z^{-n} \tag{2.58}$$

where n is an integer and z is a complex number given by $A^{j\kappa}$. The effect of this is, if we translate the expressions for downsampling, upsampling and the combination of downsampling followed by upsampling into the z-domain we get the following. The downsampling expression is given by

$$Y(z) = \frac{1}{2}[X(z^{1/2} + X(-z^{1/2})]. \tag{2.59}$$

for the upsampling expression we get

$$Y(z) = X(z^2) \tag{2.60}$$

and for downsampling followed by upsampling we get

$$Y(z) = \frac{1}{2}[X(z) + X(-z)] \tag{2.61}$$

This is because

1. Halving frequency, $\frac{\omega}{2}$ in the Fourier domain, results in $z^{1/2}$ in the z-domain.

2. Doubling frequency 2ω in the Fourier domain, results in z^2 in the z-domain.

3. Shifting frequency by π in the Fourier domain, results in $-z$ in the z-domain [2].

2.5.3 Perfect Reconstruction Filter Banks

As previously explained, a filter bank is a set of filters that are linked by sampling operators. In a two-channel filter bank, which may be seen in Figure 2.9, the analysis filters are high pass and low pass which in a specialized case are equatable to an implementation of the DWT.

It must be noted, that when performing denoising or wavelet coefficient thresholding, there is an extra stage in between of the analysis and synthesis stages. The wavelet coefficient thresholding stage suppresses all wavelet coefficients, d_j to zero if they are below or above a certain threshold. This process will be explained in more detail in a later section.

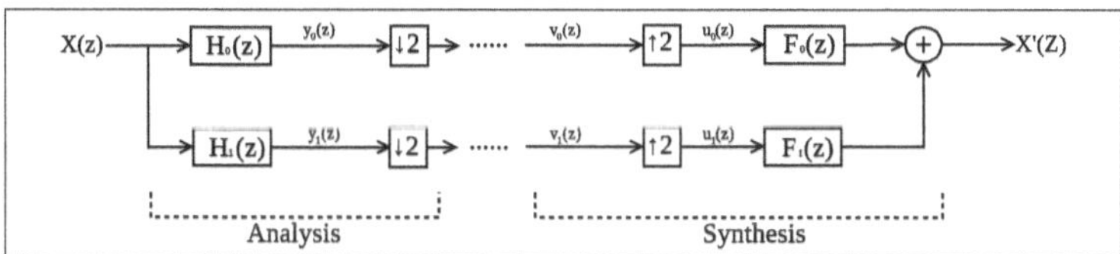

Figure 2.9: Two Stage, Two Band Perfect Reconstruction Filter Bank

As the two filters, $H_0(z)$, and, $H_1(z)$, do not have brick wall responses, implies that their frequency responses overlap. The result of this is that there is aliasing in each channel which presents itself as spectral duplication. This means that because we require perfect reconstruction for the DWT, it is necessary to adapt the synthesis filters, $F_0(z)$, and, $F_1(z)$, to cancel the spectral aliasing errors from the synthesis bank [2].

In order to derive the criteria for perfect reconstruction, the z-domain based representations of the filter bank in Figure 2.9 is considered.

Therefore if we consider

$$y_0(z) = X(z)H_0(z) \quad \text{and} \quad y_1(z) = X(z)H_1(z) \tag{2.62}$$

means using the z-domain representation in Equation (2.59) the following is obtained

$$v_0(z) = \frac{1}{2}[X(z^{1/2})H_0(z^{1/2}) + X(-z^{1/2})H_0(-z^{1/2})] \tag{2.63}$$

and

$$v_1(z) = \frac{1}{2}[X(z^{1/2})H_1(z^{1/2}) + X(-z^{1/2})H_1(-z^{1/2})]. \tag{2.64}$$

This means when these signals are then upsampled it can be seen that

$$u_0(z) = \frac{1}{2}[X(z)H_0(z)F_0(z) + X(-z)H_0(-z)F_0(z)] \tag{2.65}$$

and

$$u_1(z) = \frac{1}{2}[X(z)H_1(z)F_1(z) + X(-z)H_1(-z)F_1(z)]. \tag{2.66}$$

Finally when these two components are added together the following expression is obtained [2]

$$
\begin{aligned}
X'(z) &= \frac{1}{2}[H_0(z)F_0(z) + H_1(z)F_1(z)]X(z) \\
&+ \frac{1}{2}[H_0(-z)F_0(z) + H_1(-z)F_1(z)]X(-z)
\end{aligned}
\tag{2.67}
$$

Therefore, the two conditions for perfect reconstruction can be said to be

$$H_0(-z)F_0(z) + H_1(-z)F_1(z) = 0 \tag{2.68}$$

and

$$H_0(z)F_0(z) + H_1(z)F_1(z) = 2z^{-1} \tag{2.69}$$

which are also commonly known as the alias cancellation constraint and the no distortion constraint respectively [2].

In order to achieve the alias cancellation constraint, the following condition can be forced

$$F_0(z) = H_1(-z) \quad \text{and} \quad F_1(z) = -H_0(-z). \tag{2.70}$$

Furthermore, in order to achieve the no distortion condition, the frequency responses of the two complimentary filters must be chosen wisely. This can be done through the use of Quadrature Mirror Filters (QMF) [13]. These filters impose the following symmetry conditions on the high and low pass filters

$$H_1(z) = H_0(-z). \tag{2.71}$$

The implications of all these conditions are such that

$$h_1(n) = -(-1)^n h_0(L - 1 - n) \tag{2.72}$$

and

$$f_0(n) = -(-1)^n h_0(n) \quad \text{and} \quad f_1(n) = (-1)^n h_0(n) \tag{2.73}$$

2.6 Wavelet Coefficient Shrinkage

The technique used for interference removal with discrete wavelets is called coefficient shrinkage or thresholding. Thresholding is performed by suppressing all detail coefficients from the analysis stage of the DWT that are above or below a particular threshold. Such that, when the signal is reconstructed in the synthesis stage of the DWT, only the desired features are left intact. A block diagram illustrating this procedure can be seen in Figure 2.10.

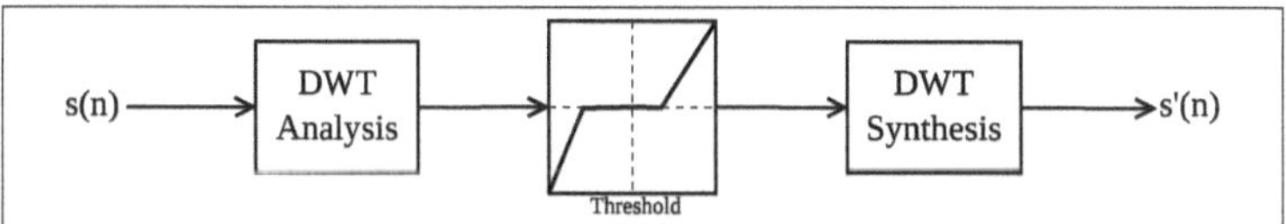

Figure 2.10: Block Diagram detailing the DWT Thresholding Procedure

The optimal choice of threshold is a statistical process that uses various estimates to ensure that the threshold is properly set. Significant research was done in this field by Donoho and Johnstone [15–17], where they describe a variety of different thresholding strategies.

A key distinction between thresholding techniques is that of level dependent and global thresholds. Level-dependent thresholding determines a new threshold for each level of detail coefficients of the decomposed signal, where as global thresholding uses a fixed threshold for each level.

2.6.1 Threshold Estimates

Universal Threshold

The universal threshold or *sqtwolog* is a global threshold based off the a measure of the noise variance. This measure of noise variance is determined by the Median Absolute Deviation (MAD) which is given by

$$\hat{\sigma} = \frac{\text{median}(|d_{J-1,k} - \text{median}(d_{J-1,k})|)}{0.6745} \tag{2.74}$$

where $\text{median}(d_{J-1,k})$ is the median along samples k of the highest resolution detail coefficients and the factor in the denominator is a factor which depends on the distribution of $d_{J-1,k}$. It is usually set to 0.6745 for scaling normally distributed data [18]. The highest detail coefficients are used for an estimate of the noise variance, as they in most cases only consist of decomposed noise [17].

The calculation of the universal threshold is determined by

$$\lambda = \hat{\sigma}\sqrt{2log(L)} \tag{2.75}$$

where L is number of samples in the signal being estimated.

Minimax Threshold

The minimax threshold uses a global threshold that is determined by minimizing the mean square error of the thresholded coefficients with an estimator. The most commonly used estimator is a least squares approach to find appropriate estimation of the noise. This threshold is referred to as minimax, as it enforces the maximal results in the worst case scenario for estimating the noise.

Stein's Unbiased Risk Estimate Threshold

Stein's Unbiased Risk Estimate (SURE) makes use of a quadratic loss function to estimate a particular threshold adaptively for each level of the decomposition. It must be noted that the SURE threshold is commonly referred to as *rigrsure* and is an adaptive threshold selection based on the use of the principle of SURE [15].

2.6.2 Wavelet Thresholding Systems

In wavelet coefficient shrinkage literature, there exist a number of ways to apply the threshold estimates to the detail coefficients of the decomposed signals [15–17]. Two of the major thresholding implementations are hard and soft thresholding.

Hard thresholding is the process that transforms all coefficients that have a magnitude below the threshold to zero, as defined by

$$\delta_\lambda^H(d_{j,k}) = \begin{cases} 0 & \text{if} \quad |d_{j,k}| \leq \lambda \\ d_{j,k} & \text{if} \quad |d_{j,k}| > \lambda \end{cases} \tag{2.76}$$

Whereas in soft thresholding, the transitions between the threshold and the function are offset by some amount, such that the transitions are smoother. This process is defined by

$$\delta_\lambda^S(d_{j,k}) = \begin{cases} 0 & \text{if} \quad |d_{j,k}| \leq \lambda \\ d_{j,k} - \lambda & \text{if} \quad d_{j,k} > \lambda \\ d_{j,k} + \lambda & \text{if} \quad d_{j,k} < -\lambda \end{cases} \tag{2.77}$$

An illustration of these thresholding definitions may be seen in Figure 2.11.

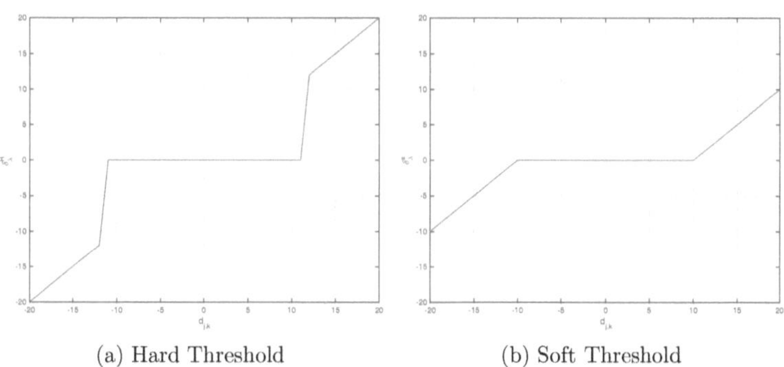

(a) Hard Threshold   (b) Soft Threshold

Figure 2.11: Figures Demonstrating the Effect of Different Types of Thresholding Techniques

2.7 Parameterization of Filter Coefficients

In order to find solutions to the dilation equation, given by Equation (2.17), it is necessary to examine the admissibility conditions for both the scaling functions, $\varphi(t)$, and the low pass filter coefficients, $h_0(n)$.

This is done in order to exploit the non-redundant representation of signals in the DWT, so that designs of optimal wavelets can be achieved in later sections. It must be noted that all proofs of these conditions may be found in a text book by Burrus *Introduction to Wavelets and Wavelet Transforms* [13]

2.7.1 Admissibility Conditions

Condition 1

If $\varphi(t) \in L^1$ is a solution to the recursive equation in (2.17) and if $\int \varphi(t) \neq 0$ then

$$\sum_n h_0(n) = \sqrt{2}. \tag{2.78}$$

Condition 2

If $\varphi(t) \in L^1$ is a solution to the recursive equation in (2.17), if $\int \varphi(t) = 1$ and

$$\sum_l \varphi(t - l) = \sum_l \varphi(l) = 1 \tag{2.79}$$

then

$$\sum_n h_0(2n) = \sum_n h_0(2n + 1). \tag{2.80}$$

This is called the fundamental condition, and is a result of requiring the recursive equation to be consistent when evaluated across all integers.

Condition 3

If $\varphi(t) \in L^2$ is a solution to the recursive equation in (2.17) and if its integer translates are orthogonal as defined by

$$\int \varphi(t)\varphi(t-k)dt = \delta_{tk} \tag{2.81}$$

then

$$\sum_n h_0(n)h_0(n-2k) = \delta_{nk}. \tag{2.82}$$

This condition shows that in order for solutions of (2.17) to be orthogonal after integer translation then it is necessary for the coefficients of the recursive equations to be orthogonal to themselves after decimation by two. It must be noted that this is another way to phase the QMF conditions discussed previously [13].

From this condition two more forms may be derived, the first being

$$\sum_n |h_0(n)|^2 = 1 \tag{2.83}$$

in order to ensure orthogonality for any normalization of $\varphi(n)$ [13]. Furthermore, it can be said that

$$\sum_n h(2n) = \sum_n h(2n+1) = \frac{1}{\sqrt{2}}. \tag{2.84}$$

In addition to the admissibility conditions mentioned above, there exist several more in the frequency domain. Notably, in work done by Daubechies [5], frequency domain constraints for developing maximally flat wavelet filters are used. These filters ensure that the frequency domain response of a particular $h_0(n)$'s are as flat as possible.

2.8   Additional Theory

In addition to the filter bank based implementations of discrete orthogonal wavelet transforms, there exist a number of alternative frameworks. As there exists literature that pertains to the suppression of interference using wavelet schemes other than the ones discussed, it is necessary to briefly describe the theory surrounding them.

2.8.1 Wavelet Lifting

An alternative to the dual-band tree structured filter-banks described in the previous section is the wavelet lifting scheme. This method was first introduced by Sweldens [19] and is a method of time-domain based interpolation broken up into three steps, the split step, the predict step and the update step.

It has been shown that the lifting scheme has superior time complexity to that of the dual-band tree implementation [3]. However as time complexity is not a constraint for this research project, all implementations of the DWT were carried out using the filter-bank based approach.

2.8.2 Biorthogonal Wavelets

Another key facet of wavelet theory is that of biorthogonal wavelets. Biorthogonal wavelets are wavelet transforms that employ a dual basis that are not orthogonal to each other for the representation of functions such that

$$g(t) = \sum_k \langle g(t), \tilde{f}_k(t) \rangle f_k(t). \tag{2.85}$$

Biorthogonal systems are used for more generalized wavelet expansions and as two basis are used biorthogonal wavelet are capable of representing a larger class of functions. However, this is at the cost of complexity of the transform, as two expansion sets need to be found, computed and stored [13].

2.8.3 Wavelet Packets

The last additional approach to be mentioned is the Wavelet Packet Transform (WPT). It is shown that rather than performing a orthonormal non-redundant expansion such as the filter bank approach of the DWT, it is possible to first represent a function as a redundant expansion and then prune the unnecessary elements to be left with the *best basis* expansion. The redundant expansion of the WPT, is performed by decomposing the wavelet space $\mathcal{W}_J$ into extra wavelet and scaling spaces $\mathcal{W}_{J+1}\ \mathcal{V}_{J+1}$ until some limit is reached, this procedure can be seen in Figure 2.12.

Once the function is expanded into redundant basis, a pruning algorithm is applied to find the optimal representation of a function. Coifman and

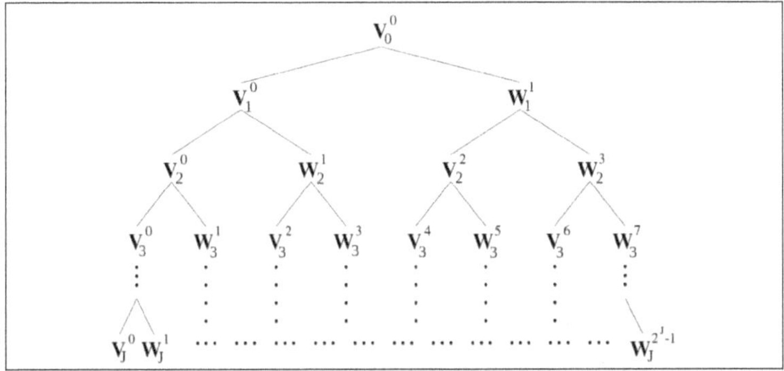

Figure 2.12: Figure Showing the Binary Tree Representation of the Wavelet Packet Transform with Redundant Basis [20]

Wickerhauser [21] propose a entropy based selection algorithm for finding the *best basis*. This is achieved by traversing the binary tree and determining the entropy at each level of the tree, such that the level of the tree with the lowest entropy is considered the optimal expansion for that resolution.

The major problem with this algorithm when applied to generalized interference suppression is that the optimal representation would have to be determined for each different class of interference. As there exist many different types of interference, it becomes impractical (in terms of time complexity) to determine the best basis. For this reason, wavelet packet decompositions are not implemented in this research.

Chapter 3

Literature Review

In order to effectively describe and design a system capable of interference suppression it is necessary to review existing literature. The literature covered in this chapter pertains to the characterization, detection and mitigation of interference as well as state of the art techniques that are used in radio astronomy. This chapter is concluded with a discussion of the literature that is used for the design of a wavelet based suppression system.

3.1 Characterization of Interference

There exist a variety of different sources of interference in microwave remote sensing and radiometry. These sources can be made up of both anthropogenic as well as natural sources. Where some examples of anthropogenic RFI sources are cell phones, satellite communications and radars whereas examples of natural sources are those of solar flares and lightening. In addition to this, distinctions can be made regarding the signatures of interference sources in both the time and frequency domains [22]. The detected signals can be described as continuous or transient in time for a particular observation window, or the observations' frequency domain characteristics can be described as narrow-band or broadband.

Due to the fact that there exist a variety of different human sources of RFI, and each source uses a different technology, there can be vast variations in the power, waveform type and frequency band of operation of the received interference signals. For this reason, the classification and characterization of interference is an immensely challenging problem.

In work by Czech et al. [23], it was shown that interference presents itself as non-Gaussian. This conclusion was achieved through the use of the Lilliefors test [24] which is a normality test that is equivalent to the Kolmogorov-Smirnov test when the mean and standard deviation is determined from the data.

3.2   Detection of Interference

In previous approaches, characterization is performed through understanding the physical and geographical phenomena for a particular observation. In work by Doran [25], three attributes were used in the classification and detection of RFI in the context of radio astronomy, these being the orientation of the telescopic array, the intensity of the RFI and the time of day a particular RFI event was measured.

In more abstracted approaches, such as that done by Lahtinen et al. [22], considerations of the polarimetry and kurtosis associated with particular RFI events were used as characterization features of RFI. Whereas other research has described methods of characterization RFI using only the time domain signal and machine learning techniques for the detection of interference [26].

This being said, each of these papers focus mainly on the classification and detection of interference, which is significantly different to the mitigation of the interference. For this reason, in order to complement the detection and classification of interference, literature must be reviewed that pertains to mitigation of interference in the context of electrical engineering.

3.3   Interference Mitigation Strategies

There are a number of techniques used for the mitigation of interference. Generally, two approaches are commonly used, namely excision and cancellation. Excision is the process of mitigating interference by clipping or blanking the temporal or spectral components of a signal that are corrupted by interference. Cancellation is the process of subtracting interference from a received signal based on an accurate classification model [27].

These techniques can be segmented into two categories of approaches used for the mitigation of interference, these being regulatory and technical methods. Regulatory approaches rely on regulatory bodies to enforce spectrum management and licensing, to ensure a high quality of service to end users of a particular frequency band [28].

In order to ensure that radio telescopes can operate at maximum sensitivity for astronomical observations, their sites are selected in remote geographical regions, which provide an environment without terrestrial interference. It can be noted that sites like these are aptly referred to as *radio quiet zones* [29].

Technical methods for the mitigation of RFI attempt to actively mitigate RFI while trying to maintain the same power level of the signal of interest. These technical methods may be segmented into three subcategories, these being RF front-end based, digital real-time suppression and offline data processing. A graphical illustration of the separation of various interference mitigation techniques can be seen in Figure 3.1.

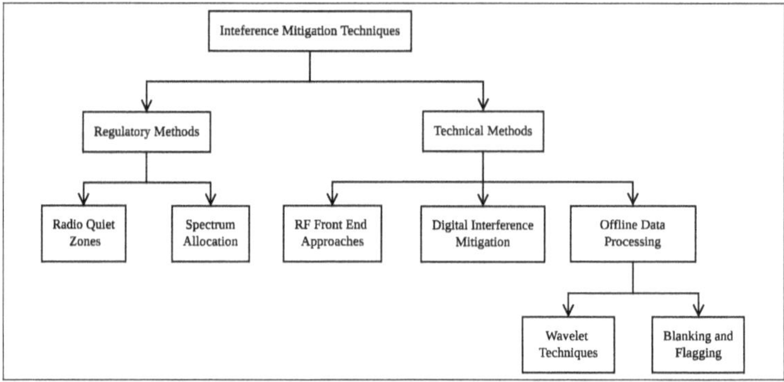

Figure 3.1: Block Diagram showing the Segmentation of RFI Mitigation Techniques

3.3.1 RF Front-End Approaches

As a first stage tactic to suppress interference, antennas are normally designed to be very directive and possess very low side-lobe levels for the rejection of strong narrow band signals. The impact of this is that any signal out of the beam pattern of a antenna is highly attenuated. Furthermore, most receivers used for radio-astronomy purposes are designed to have high dynamic ranges so that even in the presence of strong RFI signals the antennas still operate in their linear regions. In some cases, RF front ends implement automatic gain control feedback systems that adjust the gain of the receiver proportional to the received signal strength in order to avoid saturation [28].

In passive remote sensing a reference antenna is often used to actively cancel RFI. This is a achieved by using adaptive filters that filter the surveillance signal and change its filter weights depending on the RFI detected on the reference channel [30].

These methods are effective, but as spectrum becomes more crowded the RFI becomes less predictable, and more flexible techniques are required.

3.3.2 Digital Interference Mitigation

Digital interference mitigation algorithms have a variety of implementation strategies. An effective distinction to be made when implementing a digital interference mitigation algorithm is whether the process must be implemented in real-time or on offline processors when the results are not required immediately. The real-time or offline distinction allows for the determination of which digital system technology will be used for the application of the algorithm of choice.

Most commonly, Field Programmable Gate Arrays (FPGAs) or Graphical Processing Units (GPUs) are used for real time implementations. This is done in order to increase an algorithm's throughput and speed through the use of parallelization.

In addition to the considerations previously mentioned, digital interference suppression algorithms are often classified by the domain in which their operations are performed. Generally, there are three domains where interference suppression algorithms can be applied, the temporal, spectral and spacial domains [28].

Time Domain Thresholding

A common way of removing transient interference is using thresholding techniques in the time domain. This is achieved through the use of temporal blanking when the received signal exceeds a threshold, or by replacing the detected interference with a particular signal. The thresholds which govern this class of mitigation algorithms are defined by what particular interference is present in a system. For example in a study done for the Giant Metrewave Radio Telescope (GMRT) [31], interference transient signals were thresholded effectively using the Median Absolute Deviation (MAD) estimator. Whereas other approaches such as the VarThreshold using Cumulative Sum (CUSUM) have been shown to be effective [32].

Adaptive Filtering

For real-time flexible RFI cancellation, least-mean-squared (LMS) adaptive filters are often used in remote sensing. In a paper by Lord et al. [33] a design and validation of a LMS filter algorithm is described. The LMS filter is designed using a feedback loop and a digital filter that tries to minimize the error between the input and output by recursively altering the filter taps weights at each sampling instant until the filter taps converge to some static value. In this approach, the converged filter's transfer function is then used for the preceding received signals, thereby increasing time efficiency of the algorithm.

Spacial Filtering

Spacial filtering is a digital interference mitigation strategy that removes interference that is localized in a particular direction by forming a null in the beam pattern. This method is only applicable for antenna arrays with electronically steerable beams [28].

3.3.3 Offline Data Processing

In the case of radio astronomy it is possible to perform post processing in order to suppress interference. Offline processing is preferred technique in astronomy due to high data volumes over long observation times. These methods are often categorized into flagging and blanking techniques.

Flagging is a method of interference mitigation that suppresses interference corrupted data after correlation whereas blanking achieves the same results however it acts on the pre-correlated data [34].

Flagging can be performed either manually or automatically, where manual flagging requires astronomers to sift through the large data sets and assign labels in accordance to their experience. Whereas automated flagging algorithms such as *AOFlagger* use statistical methods in the time and frequency domains to determine corrupted regions in a measurement set [35].

Even though flagging is suitable to interference excision in the radio astronomical context, it is a lossy suppression technique. The result of which is that if a lower power observation is embedded below an RFI event, then flagging will remove those observations as well. This is destructive to measured data and more appropriate techniques are investigated in this research project.

3.4 Wavelet Suppression Schemes

Due to the fact that interference presents its self as non-stationary in both the time and frequency domains, the classical techniques mentioned above are often not well suited for interference excision. Wavelet decompositions offer good time and scale localizations thereby being more suitable for detecting non-stationary signals such as RFI. The application specific details of interference excision approaches using wavelets and wavelet theory will be described in later sections.

3.4.1 Wavelet Based Excision

Wavelet based interference excision forms a subsection of the research field of wavelet based denoising algorithms. Wavelet based denoising or coefficient shrinkage is the process of thresholding each scale of the decomposed signal and re-synthesizing the thresholded coefficients to form the denoised signal through the inverse wavelet transform. In a seminal paper by Donoho and Johnstone [17] it is shown that there exist a number of thresholds that can be applied to the a signal using either soft or hard thresholding. Where hard thresholding sets all coefficients at a particular scale that are below the threshold to zero, whereas soft thresholding offsets the threshold by the coefficient's particular index at that scale.

In work on RFI excision for radio astronomy done by Maslaković et al. [36], it is shown that when interference power levels far exceed the astronomical observations we can invert the thresholding such that only the RFI is reconstructed. Through this, it is shown that if one subtracts the reconstructed RFI from the original signal then the astronomical observation is recovered. Furthermore, in order to maximize the performance of the excision algorithm it is shown that if the wavelet used for the decomposition portrays similar temporal qualities to the RFI, then optimal excision is achieved.

3.4.2 Wavelet Design

In order to maximize performance for non-stationary interference excision it is necessary to design suitable wavelets for detection. There are several approaches for designing wavelets that offer compact representations of interference signals. These design strategies can be separated into two different classes, those for library design and those for single wavelet design.

In a paper by Chourasia [37] a design methodology for new wavelet basis for phonocardiography is described. This is achieved by using a parametrization of the PR filter bank and ensuring that the filter taps result in features similar to that of the phonocardiographic signal of interest. A problem with this approach is that the design process is quite laborious for the design of a single wavelet function, and that for the purposes of general interference excision it is not the most suitable.

In a seminal paper by Chapa and Rao [38], they describe algorithms for designing specific orthogonal wavelets to match specific signals. This is done by minimizing the error between the spectra of the wavelet and the function of interest. Furthermore, their research is extended to the discrete domain such that impulse responses of the QMF filters can be determined. Similarly to the previous approach the specified algorithm is tedious for the design of a single wavelet, and as previously mentioned for the purposes of general interference excision it is non-ideal.

In works by Ong [39], a technique using genetic algorithms to optimize trigonometrically parameterized wavelet filters is described. The cost function associated with the genetic algorithm based optimization is a measure of Shannon's entropy [21]. This optimization is performed by taking the discrete wavelet decomposition of a linear swept sinusoid and analyzing the entropy of the waveform at each scale of the decomposition and through the use of simulated annealing the best basis is found.

In addition to the algorithms mentioned, there exists research [40, 41] for designing discrete wavelets for only the analysis stage of the filter bank representation. This gives more flexibility for wavelet designs as some of the admissibility conditions may be dismissed, thereby having less constraints on the optimization procedure. These techniques are have been used for characterization and detection of signals in the wavelet domain for the purposes of image processing and fault detection of systems. However, as these techniques do not ensure that the synthesis filter's in the PR filter-bank approach exist, they are not suitable for RFI excision and therefore cannot be used.

In work done by Maslaković et al. [42], a library based approach to wavelet design is documented. This approach is useful as it characterizes a library of wavelets suitable for RFI detection and excision by performing a multi-level minimization procedure. The multi-level procedure applies two thresholds to the wavelet search space, the first is smoothness and the second is similarity. The smoothness threshold ensures that the resulting interference-matched wavelets in the library have a second order difference equation below some predetermined value, and the similarity threshold ensures that none of the smooth wavelets are similar. The details of this algorithm are discussed in Chapter 4.

3.4.3 Matching Wavelets to RFI

In research done by Coifman and Wickerhauser [21] an adaption of Shannon's Entropy was used to determine the optimal basis for a wavelet-packet decomposition. This approach found the most compact basis representation of a signal at a particular scale in the DWT. This algorithm was misleadingly named the *best basis* algorithm as it selects the most suitable binary tree for a given wavelet packet representation. However in more standard implementations of the DWT the *best basis* algorithm does not often result in the best basis.

Maslaković et al. [43] showed that there are more suitable techniques to match wavelets to RFI for optimal RFI excision. The optimal time shift algorithm and Fractional Excised Energy (FEE) measure are described in this paper. Where the optimal time shift algorithm addresses the time-variance property of the DWT which may not lead to optimal energy of each level of the DWT if the input signal is shifted in samples. This is done by searching a binary tree representing the energy costs associated with a shifted input signal, and traversing it to find the optimal shift. Whereas the FEE shows the ratio of energy at each scale of the RFI in relation to the orignal signal.

Chapter 4

Smooth Orthonormal Wavelet Library Design

This chapter shows the formulation and optimization of the design of orthogonal wavelet libraries. In order to create wavelets that match to various sources of interference it is necessary to consider their formulation. This is achieved through the use of the Strict Sequence Decomposition (SSD) parameterization of wavelet filter coefficients and optimizing the resultant wavelets for smoothness and uniqueness. This chapter is concluded with a multi-dimensional wavelet library that is well suited to interference suppression.

4.1   Introduction to Wavelet Design

As described in the previous chapter, wavelet based techniques are well suited to interference suppression in the context of radio astronomy as interference often presents itself as non-stationary. This non-stationary characteristic makes Fourier based techniques ill-equipped for the purpose of excising interference from the observed signals.

In order to achieve interference suppression using wavelets, it is imperative to determine a wavelet that best matches the interference that has corrupted the recorded signal. A wavelet that is well suited to interference excision has the most compact decomposition in the wavelet domain, leading to the optimal suppression of unwanted interference. This being said, the optimal wavelet for interference suppression must also preserve the underlying features of the astronomical signal [43].

Wavelet literature often makes use of common families of wavelets that possess some particular useful characteristic. For example Daubechies wavelets ensure the maximum number of vanishing moments resulting in the maximum compactness of support, whereas Coiflets ensure that both the wavelet and scaling function share the same number of vanishing moments [5]. However for the purposes of interference excision it is often non-optimal to use existing families of wavelets as they do not offer the most compact representations of the interference to be excised.

This being said, methods do exist for designing signal-matched wavelets, however they have been shown to be computationally-expensive and make use of complicated optimization procedures that only result in a single wavelet [38, 44–46]. Therefore, as this implementation would like to have a comprehensive library for the excision of interference, it is useful to construct a general library of smooth orthogonal wavelets. For the purpose of interference suppression a wavelet library must be designed that ensures that its entries are orthogonal, smooth and unique.

The reason why smoothness is an important characteristic of the designed wavelets is due to the fact that the chosen wavelet both impacts the compactness of the representation in the wavelet domain as well as the shape of the thresholded signal. This means that if a non-smooth wavelet is used for a decomposition of a particular interference corrupted signal and it is suppressed using coefficient shrinkage resulting in the signal being resynthesized. Then artefacts from the choice of wavelet will appear in the resynthesis of the interference-suppressed signal. This means that if the chosen wavelet has harsh discontinuities, then the resynthesized signal will also contain remnants of the non-smooth wavelet.

For this reason, smoothness is an important characteristic of the library of wavelets to be designed. The design process of this library is documented in this section.

4.2 Discrete Wavelet Systems

As per Mallat's Algorithm [47] the DWT is implemented using a perfect reconstruction filter-bank approach. The effect of this is the DWT decomposition can be implemented through the nesting of low-pass and high-pass filters as shown by Figure 4.1. Where the high-pass filters are defined by, h_0, and are referred to as the wavelet sequences, whereas, h_1, are the low-pass filters and are referred to as the scaling sequences.

The consequence of this system is that the outputs of the high-pass filters, d_J, result in the input signal being analyzed by a wavelet function at a particular scale and are referred to as the detail coefficients. Similarly the outputs of the low-pass filters, c_J, are the result of the input signal being analyzed by a scaling function at a particular scale and are referred to as the approximation coefficients.

As explained in Section 2.7, there exist various admissibility criteria that determine when a particular set of filter coefficients may be used as wavelet sequences. From these constraints it is possible to design a wavelet library with particular useful characteristics for interference excision.

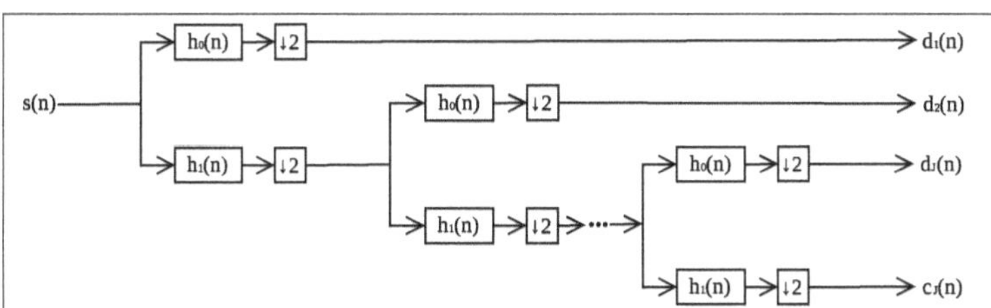

Figure 4.1: DWT Analysis Bank

4.2.1 Parametrization of Wavelet Filter Coefficients

In order to create search space in which to look for particular wavelets that are suited to an interference suppression wavelet library, it is necessary to parameterize the the admissibility conditions. The parameterization is performed in decompositions of $\sin(\alpha + \beta)$ and $\cos(\alpha + \beta)$ and is referred to as Strict Sequence Decomposition (SSD) [48].

According to the SSD rules any set of parameters $\{\alpha_1, \alpha_2, \ldots, \alpha_k\}$ in the form of $\sin(\alpha_1 + \alpha_2 + \cdots + \alpha_k)$ and $\cos(\alpha_1 + \alpha_2 + \cdots + \alpha_k)$ can be used to describe a wavelet filter if their sum is equal to $\pi/4$ [49]. Through these methods the various decomposition length sequences shown below.

Length-2 Scaling Sequences

In the case that $n = 1$, the scaling sequence has two coefficients as shown by

$$h_0(0) = \cos(\alpha) \quad , \quad h_0(1) = \sin(\alpha). \tag{4.1}$$

Length-4 Scaling Sequences

In the case that $n = 2$, the scaling sequence has four coefficients as shown by

$$\begin{aligned}
h_0(0) &= \cos(\alpha)\cos(\beta) \quad , \quad h_0(2) = -\sin(\alpha)\sin(\beta) \\
h_0(1) &= \sin(\alpha)\cos(\beta) \quad , \quad h_0(3) = \cos(\alpha)\sin(\beta)
\end{aligned} \tag{4.2}$$

where $\beta = \pi/4 - \alpha$

Length-6 Scaling Sequences

In the case that $n = 3$, the scaling sequence has four coefficients as shown by

$$\begin{aligned}
h_0(0) &= \cos(\alpha)\cos(\beta)\cos(\gamma) \quad , \quad h_0(3) = \cos(\alpha + \gamma)\sin(\beta) \\
h_0(1) &= \sin(\alpha)\cos(\beta)\cos(\gamma) \quad , \quad h_0(4) = -\sin(\alpha)\cos(\beta)\sin(\gamma) \\
h_0(2) &= -\sin(\alpha + \gamma)\cos(\beta) \quad , \quad h_0(5) = \sin(\alpha)\cos(\beta)\cos(\gamma)
\end{aligned} \tag{4.3}$$

where $\gamma = \pi/4 - \alpha - \beta$

Length-16 Scaling Sequences

In the case that $n = 4$, the scaling sequence has four coefficients as shown by

$$\begin{aligned}
h_0(0) &= \cos(\alpha)\cos(\beta)\cos(\gamma)\cos(\delta) \quad , \quad h_0(8) = -\sin(\alpha)\sin(\beta)\cos(\gamma)\cos(\delta) \\
h_0(1) &= \sin(\alpha)\cos(\beta)\cos(\gamma)\cos(\delta) \quad , \quad h_0(9) = \cos(\alpha)\sin(\beta)\cos(\gamma)\cos(\delta) \\
h_0(2) &= -\cos(\alpha)\cos(\beta)\sin(\gamma)\sin(\delta) \quad , \quad h_0(10) = \sin(\alpha)\sin(\beta)\sin(\gamma)\sin(\delta) \\
h_0(3) &= -\sin(\alpha)\cos(\beta)\sin(\gamma)\sin(\delta) \quad , \quad h_0(11) = -\cos(\alpha)\sin(\beta)\sin(\gamma)\sin(\delta) \\
h_0(4) &= -\cos(\alpha)\sin(\beta)\sin(\gamma)\cos(\delta) \quad , \quad h_0(12) = -\sin(\alpha)\cos(\beta)\sin(\gamma)\cos(\delta) \\
h_0(5) &= -\sin(\alpha)\sin(\beta)\sin(\gamma)\cos(\delta) \quad , \quad h_0(13) = \cos(\alpha)\cos(\beta)\sin(\gamma)\cos(\delta) \\
h_0(6) &= -\cos(\alpha)\sin(\beta)\cos(\gamma)\sin(\delta) \quad , \quad h_0(14) = -\sin(\alpha)\cos(\beta)\cos(\gamma)\sin(\delta) \\
h_0(7) &= -\sin(\alpha)\sin(\beta)\cos(\gamma)\sin(\delta) \quad , \quad h_0(15) = \cos(\alpha)\cos(\beta)\cos(\gamma)\sin(\delta)
\end{aligned} \tag{4.4}$$

where $\delta = \pi/4 - \alpha - \beta - \gamma$

4.2.2 Wavelet Smoothness Criteria

As the parameterization of the scaling function filter coefficients are
constrained by the admissibility conditions, the resulting wavelet are not
guaranteed to be smooth. This is because, only the linearity condition given
by Equation (2.78) ensures that only the zeroth vanishing moment is zero [36].

Wavelet smoothness prevents matching to spurious signals thereby
enabling the suppression of interference signals in many cases. Therefore,
it is necessary to ensure all wavelets in the library are confined to some
degree of smoothness. In the continuous time case, wavelet smoothness
is equivalent to regularity, which is related to the number of vanishing
moments a wavelet has and can be measured using the Hölder or Sobolev
exponents [50].

In the discrete time case, constraints may be placed on the parameterization
angles to ensure that higher order moments are zero, however this was
shown to not ensure wavelet smoothness [51]. This being said, wavelet and
scaling filter coefficients can be used to reconstruct their corresponding
wavelet, $\psi(t)$, and scaling, $\varphi(t)$ functions using the successive approximation
or cascade algorithm. Thereby enabling the use of the discrete time measure
of smoothness of the reconstructed wavelets using local variation [42].

Although the wavelet and scaling functions are never explicitly used
for analysis or synthesis, they are invaluable for measuring smoothness. In
order to solve the recursive equation given in Equation (2.17), an iterative
algorithm that generates successive approximations to $\varphi(t)$ is used, as shown
by [13]

$$\varphi^{(J+1)}(t) = \sum_{n=0}^{N-1} h_0(n)\varphi^{(J)}(2t - n). \tag{4.5}$$

As explained in Section 2.5, these operations may be implemented using
sub-samplers and convolution given that on the first iteration, $\varphi(2t - n)$ is
the up-sampled version of $h_0(n)$. Furthermore, the wavelet function $\psi(t)$ is
calculated by

$$\psi^{(J)}(t) = \sum_{k} h_1(k)\varphi^{(J)}(2t - k). \tag{4.6}$$

The mth-order local variation of a particular wavelet sequence, $\varphi(n)$, is given by the L-1 norm of the difference equation obtained from the discrete finite variation [42]. Where the difference equation of the discrete finite variation is given by

$$V^{(m)} = \sum_n \sum_{k=0}^{m} (-1)^k \binom{m}{k} \varphi(n - k). \tag{4.7}$$

From this definition it is possible to define the first-order and second-order local variation. Where the first-order discrete finite variation is given by the sum of the difference between successive samples of the wavelet sequence as shown by

$$V^{(1)} = \sum_n |\varphi(n) - \varphi(n - 1)| \tag{4.8}$$

and the second-order is given by

$$V^{(2)} = \sum_n |\varphi(n) - 2\varphi(n - 1) + \varphi(n - 2)|. \tag{4.9}$$

When trying to assemble a wavelet library it is necessary to determine the optimal local variation order. In order to analytically find the optimal order, 6 wavelets were chosen to test the local variation order. Where each wavelet was deemed to have significantly different smoothness through visual inspection.

Wavelets one to five were constructed using the parameterization for length-4 scaling sequences using angle, α, and wavelet number 6 was made to be the Daubechies 10 wavelet. These wavelets may be seen in Figure 4.2 and in Figure 4.3 each wavelet's corresponding local variation at different orders may be seen.

The first observation that can be made from Figure 4.3 is that the first two wavelets are not smooth, this is confirmed from the results shown in Figure 4.3. It can be noted that as the order increases, the more sensitive the local variation becomes, this is shown by how the distance between each of the lines corresponding to each of the wavelets increases.

Furthermore, it can be seen that the 1st order local variation does not accurately represent the smoothness of the wavelets in question. This is shown in Figure 4.3, as wavelet number 6 has a higher first-order local variation than wavelet 5, whereas from local variation levels of 2 and upward

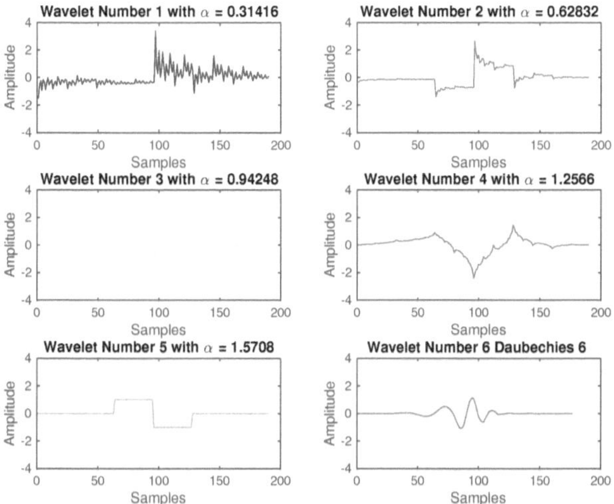

Figure 4.2: Wavelets Chosen to Determine Optimal Local Variation Order

wavelet number 6 clearly has a significantly lower local variation than wavelet 5.

It could be said that the highest order local variation level should be used as it offers the greatest sensitivity, however as the local variation order increases, the computation complexity increases exponentially. The effect of this is that when constructing smooth wavelet libraries that require the assessment of the smoothness of billions of wavelets it is necessary to keep computational complexity as low as possible.

For these reasons it was determined that the second order local variation was deemed most suitable, when considering computational complexity and sensitivity.

In order to effectively use the principals discussed surrounding wavelet smoothness it is necessary to determine a smoothness threshold. In this case, the threshold should prevent wavelets that are unsuitably smooth from being added to the wavelet library. In this case the smoothness threshold, V_H, was determined by the second-order local variation of a Haar wavelet. Therefore all wavelets that have a second-order local variation, V, less than

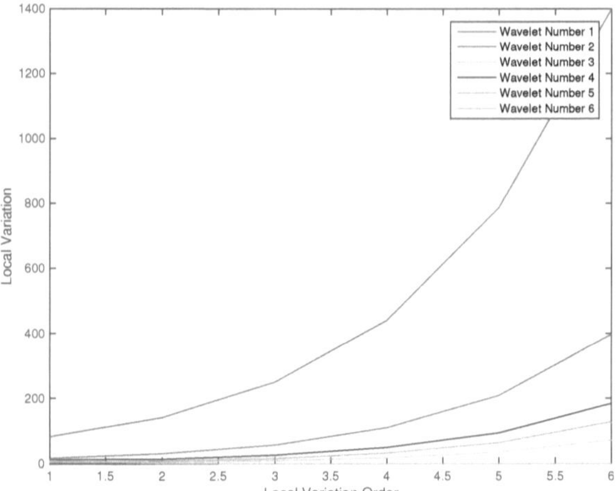

Figure 4.3: Local Variation Order vs. Local Variation of Specific Wavelets, where the Higher the order the Greater the Sensitivity

or equal to 8 are considered suitable.

4.2.3 Wavelet Similarity

As the library to be created should be contain unique smooth wavelets for interference excision, it is necessary to ensure that each of the wavelets share some degree of uniqueness. This is crucial because when in use, the wavelet library should contain a single wavelet that represents a particular interference waveform.

This is achieved by the use of the inner product, which performs the sum of the piecewise multiplication of two signals as shown by

$$\tau = \langle\, s_1(n), s_2(n) \rangle = \sum_n s_1(n)s_2(n) \leq \tau_{ip} \tag{4.10}$$

where the result of this operation is between 0 and 1 given that the two input signals are normalized.

In effect this enables a threshold to be set, that limits the similarity between the inner product of two wavelets. In order to determine the optimal threshold, trials using 8 length-4 wavelet sequences were used.

Figure 4.4a, shows a library of 8 smooth wavelets that are not unique. In order to select the wavelets that are not similar to each other, the inner product between every wavelet was taken. A similarity threshold, τ_{ip}, was then applied and was varied from 0.1 to 1 in steps of 0.05 in order to determine the optimal similarity threshold. Figure 4.4b shows the result of

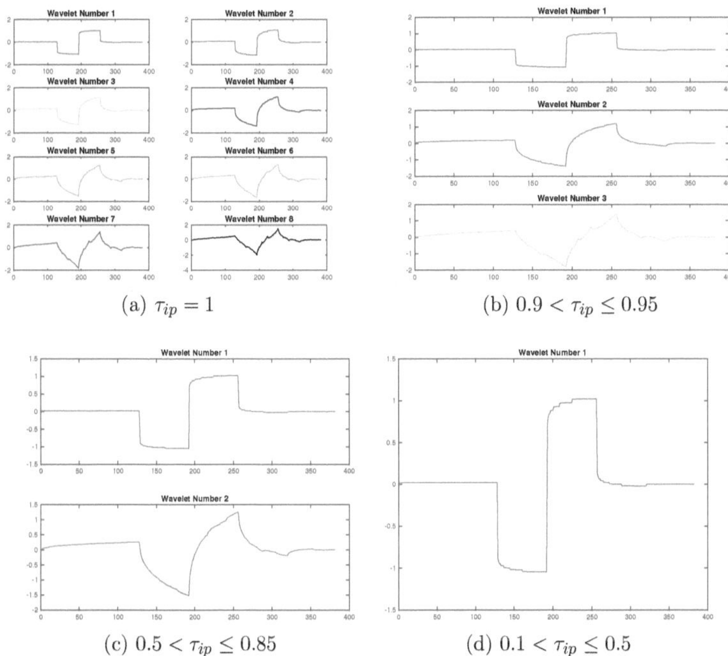

(a) $\tau_{ip} = 1$

(b) $0.9 < \tau_{ip} \leq 0.95$

(c) $0.5 < \tau_{ip} \leq 0.85$

(d) $0.1 < \tau_{ip} \leq 0.5$

Figure 4.4: Figures Showing the Effect on the Wavelet Library when Varying τ_{ip}

setting the similarity threshold between 0.95 and 0.9. It can be noted that at this threshold level the number of wavelets shrank in size from 8 to 3. Furthermore, it can be seen that all wavelets that are visually similar have been eliminated, and only unique wavelets remain.

Figure 4.4c shows the result of setting τ_{ip} less than 0.9 and more than 0.5. The effect of this, is that wavelet 3 is removed from the library, even though it is visually unique. Finally, Figure 4.4d shows the effect of setting τ_{ip} less than 0.45 and 0.1. It can be noted that when the threshold is low, only a single wavelet is included in the library, which in this case is not useful, as it is required to design a general wavelet library for interference excision. For this reason, a threshold of 0.95 was considered to be optimal, as it enabled the inclusion of dissimilar wavelets.

4.3 Populating Wavelet Libraries

As discussed previously, this project attempts to construct a general wavelet library for the purposes of interference suppression. Therefore it is necessary to include wavelets that may not seem suitable for other applications, such as wavelets with discontinuities [42]. As shown in Section 4.2.2, the smoothness threshold of $V_H \leq 8$ was deemed suitable. Furthermore the similarity threshold, τ_{ip}, was fixed to 0.95.

Table I shows the parameters used for determining suitable wavelets for the general interference excision library. From these specifications on the constructed wavelets it is possible to design an algorithm to populate the desired wavelet libraries.

TABLE I

Parameters Available to Design Optimal Wavelet

Parameter	Description	Range	Effect on Wavelet
Δs	Step Size	$(0, \pi)$	Granularity of Search
α	Parametric Angle	$[0, \pi]$	Shape of Wavelet
β	Parametric Angle	$[0, \pi]$	Shape of Wavelet
γ	Parametric Angle	$[0, \pi]$	Shape of Wavelet
δ	Parametric Angle	$[0, \pi]$	Shape of Wavelet
V_H	Smoothness Threshold	$\mathbb{Z}$	Smoothness of Wavelet
τ_{ip}	Similarity Threshold	$\mathbb{Z}$	Similarity between Wavelets
n	DWT Depth	$\mathbb{Z}$	Number of Levels in Expansion
H	Entropy	$\mathbb{Z}$	Compactness of Decomposition

In order to find the optimal wavelet library using the trigonometric parameterization of wavelet filter coefficients an algorithm was designed to populate the smooth, unique, orthogonal wavelet library for interference suppression. A flow chart representing the procedure of this algorithm may be seen in Figure 4.5.

The algorithm begins with initializing the library parameters, these being, the length of the wavelet sequence parameterization (which determine how many angles are used), the step size, Δs, and the DWT decomposition depth, n.

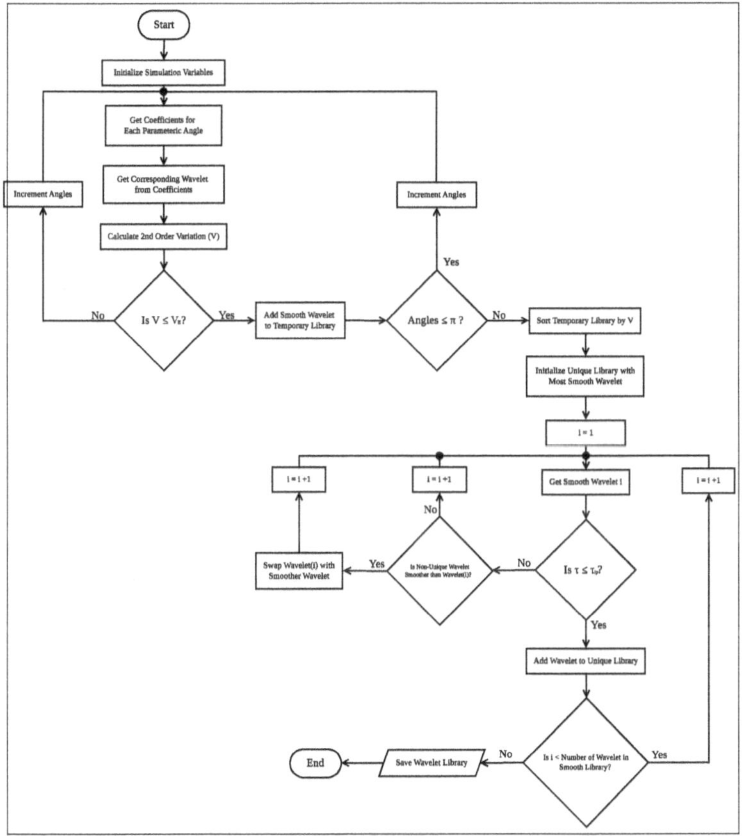

Figure 4.5: Flow Chart showing the Library Population Algorithm

Then each angle parameter is swept linearly in steps of Δs and every corresponding wavelet and second-order local variation, V, is determined from the parameterized wavelet sequence. After which each wavelet's second-order local variation is compared to the threshold, $V_H = 8$. If the currently indexed wavelet's second-order local variation is below V_H then it is added to a temporary wavelet library, otherwise it is discarded.

Once the temporary library of smooth wavelets is populated, it is sorted by V in ascending order and the most smooth wavelet is added to a smooth, unique output library. Then the temporary smooth library is iterated over and each wavelet is compared to every wavelet in the smooth, unique output library. If the two wavelets being compared are not similar ($\tau < \tau_{ip}$), then the smooth wavelet is added to the output library. If the two wavelets are similar ($\tau \geq \tau_{ip}$), then the smoother of the two wavelets is swapped into the output library.

For example, on the first iteration of the output wavelet library generator, only one wavelet is included in the output library, which in this case, is the most smooth wavelet. Then the second smoothest wavelet is chosen from the smooth library and is compared to every entry in output library, if they are similar then the output library remains unchanged. However, if the second smoothest wavelet is non-similar to the wavelet in the output library, then it is added. This procedure continues for every wavelet in the smooth library.

Using the wavelet population algorithm described above, simulations were run for each of the n-length wavelet sequence parameterizations.

4.3.1   One-Dimensional Parameter Space

A length-4 sequence generates a one dimensional parameter search space. This is because, when we vary the single parameter angle, α, there is a one-to-one mapping to a wavelet sequence. Therefore, it is possible and very useful to show the relationship between the local variation of each wavelet that corresponds to a single angle. This relationship can be seen in Figure 4.6.

In order to perform the library creation algorithm, Δs was set to 256 and n was set to 5. In Figure 4.6a the blue diamonds present on the graph correspond to a single wavelet's second-order local variation. The orange line represents the smoothness threshold V_H, which is equal to 8 and the red

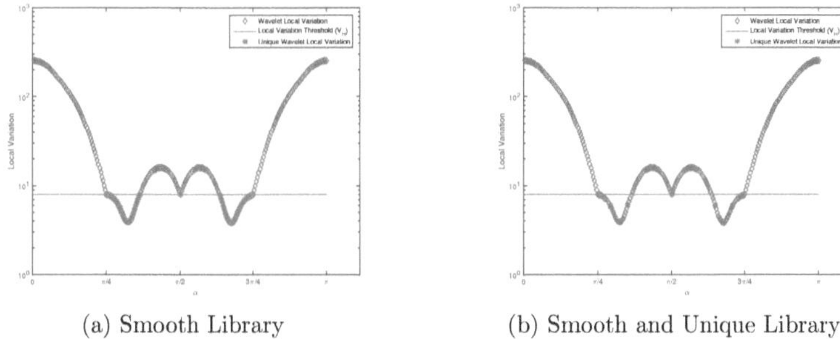

(a) Smooth Library   (b) Smooth and Unique Library

Figure 4.6: Figures Showing the Relationship between parameter angle, α, and the second-order local variation of a Wavelet Library

stars represent all wavelets that have a local variation below or equal to V_H.

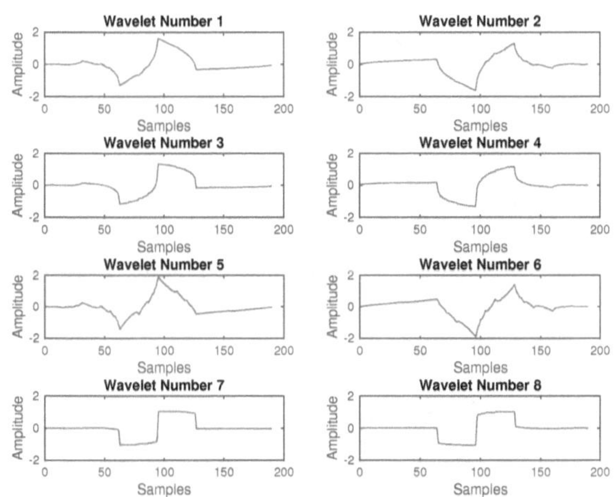

Figure 4.7: Unique, Smooth Wavelets Belonging to the 1D Search Space

Figure 4.6b shows the result of when the similarity threshold, τ_{ip} is applied to the smooth wavelet library. It can be noted, that there is an enormous reduction in the number of wavelets, between the smooth library shown by Figure 4.6a and the unique library shown by Figure 4.6b. This result proves that the library population algorithm is effective in its selection of smooth, unique wavelets.

53

It can be seen in both Figures 4.6a and 4.6b that there is a clear symmetry around the point where the parameter angle α is $\pi/2$. This phenomenon is attributable to the resultant wavelets of the length-4 parameterization being symmetrical as can be seen in Figure 4.7. The effect of this is that wavelets $\psi(t)$ are considered dissimilar to $-\psi(-t)$. It is important to keep the reversed wavelets in the outputted library as they lead to unique analysis in the wavelet domain.

Figure 4.7 shows the resultant smooth, unique wavelets that correspond to the red crosses in Figure 4.6b.

4.3.2   Two-Dimensional Parameter Space

Using the length-6 parameterization, a two-dimensional parameter search space is generated, which may be seen in Figure 4.8. This space also used, $\Delta s = \pi/256$ and $n = 5$.

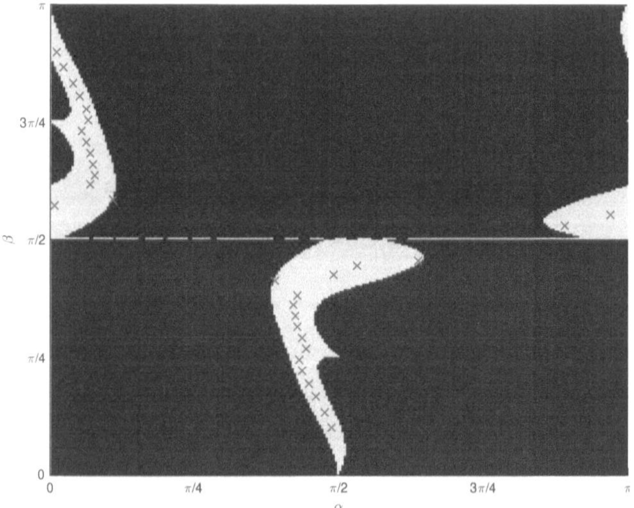

Figure 4.8: Figure showing the relationship between parameter angles, α and β, and the second-order local variation of a wavelet library. Where the blue regions are incompatible wavelets, the yellow regions are smooth wavelets and the red crosses are smooth unique wavelets.

It can be noted that the yellow regions in Figure 4.8 represents all wavelets that posses a second-order local variation below $V_H = 8$. Whereas the blue regions represent all wavelets with a second-order local variation above the threshold. The crosses within the smooth search space represents all wavelets that are smooth and unique.

It can be similarly noted that a diagonal symmetry exists in the 2D search space, this again results in wavelets $\psi(t)$ and $-\psi(-t)$ being considered unique to each other.

The resulting unique, smooth wavelets from this search space may be seen in Figure 4.9. It can be seen that there are more wavelets in the 2D search space, this may be a result of the 2D search space being exponentially larger. For the 1D search space consisted of 256 wavelets, the 2D space had 256^2. Furthermore, it can be noted that the wavelets in the 2D library appear to be smoother than the 1D library, as well as exhibiting more zero crossings.

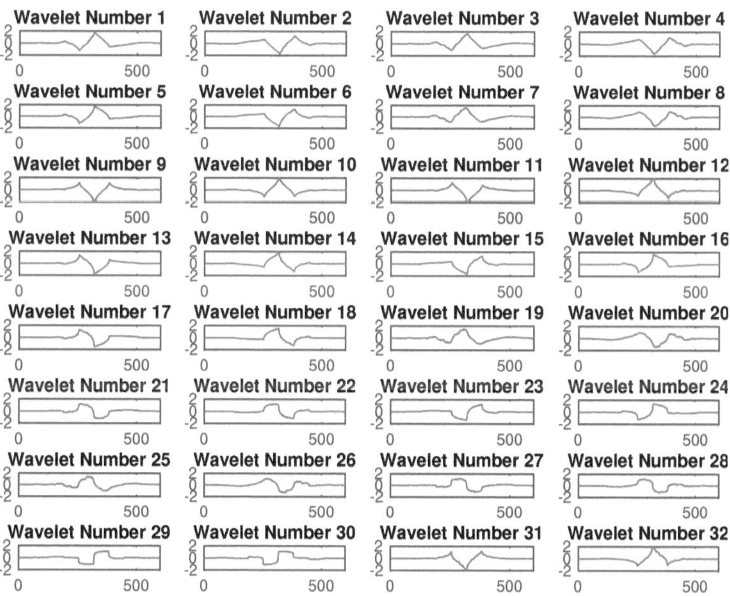

Figure 4.9: Unique, Smooth Wavelets Belonging to the 2D Wavelet Library

4.3.3 Three-Dimensional Parameter Space

Using the length-16 parameterization a three-dimensional search space is generated. As can be seen in Figure 4.10, the 3D search space is made up of the three parametrization angles α, β and γ with the colored circles representing the second-order local variation of each wavelet in the space.

Figure 4.10a shows all wavelets that posses a smoothness above the threshold $V_H = 8$, it can be noted there exists some degree of symmetry which follows the same reasons as previously mentioned. Figure 4.10b, shows the resultant unique, smooth library of wavelets that ensure that the normalized similarity between them is less that 0.95.

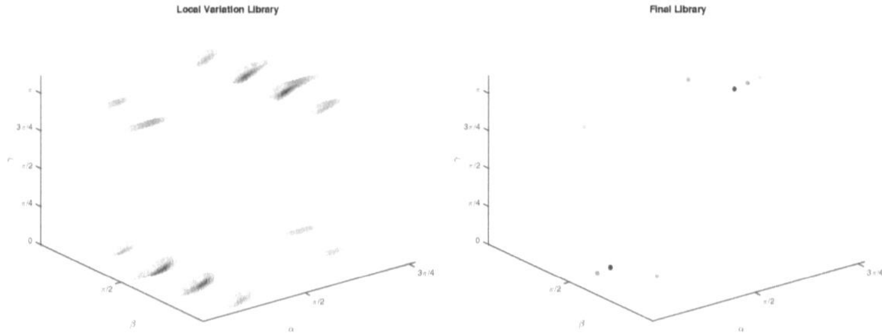

(a) A Smooth Wavelet Library   (b) A Unique, Smooth Wavelet Library

Figure 4.10: Figures Showing the Relationship between parameter angles, α, β and γ, and the second-order local variation of a Wavelet Library

In Figure 4.11 the resultant wavelets of the three-dimensional population algorithm can be seen. It can be noted that some of the wavelets appear to have more zero-crossings than both the 1D and 2D libraries. Furthermore, the resultant wavelets from the 3D parameterization seem to be smoother than the wavelets in the other libraries.

4.3.4 Multidimensional Library Comparison

In order to determine the most suitable library out of the multidimensional search spaces generated, it is necessary to compare particular attributes from each of the libraries. It must be stressed, that the desired outcome from the multidimensional wavelet library design is a collection of wavelets well

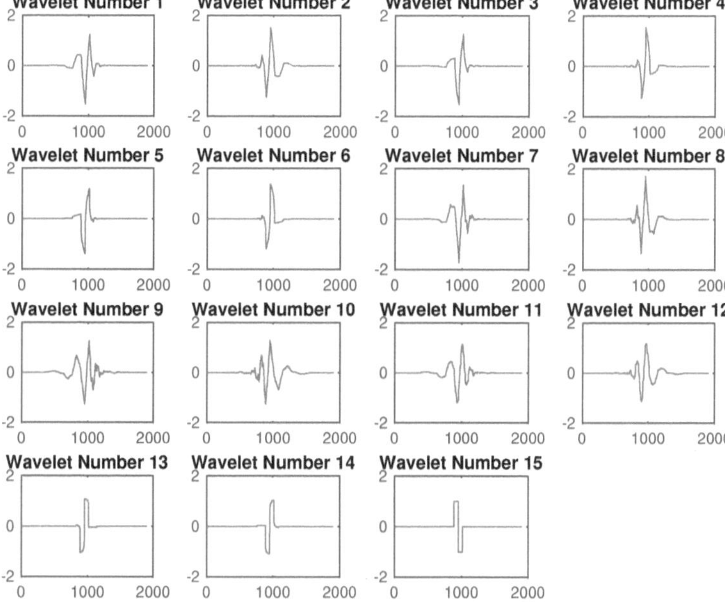

Figure 4.11: Unique, Smooth Wavelets Belonging to the 3D Wavelet Library

suited to interference suppression. Therefore, for reasons mentioned above the smoothness criterion is considered to be the most critical.

Optimal Step Size

In Figure 4.12 the relationship between the second-order local variation of each of the wavelets in every library is seen when varying the step size (Δs). It can be seen that in all three dimensions the smoothness seems to approach some limit.

In Figure 4.12, we can see the average local variation per step size of each of the wavelet libraries. It is clear that the 2D wavelet library posses the smoothest wavelets. Furthermore, this result confirms that the smoothness of wavelets decays exponentially with step size. The impact of this is that the Δs can definitively be set to $\pi/1024$ in order to ensure the optimally smooth wavelets.

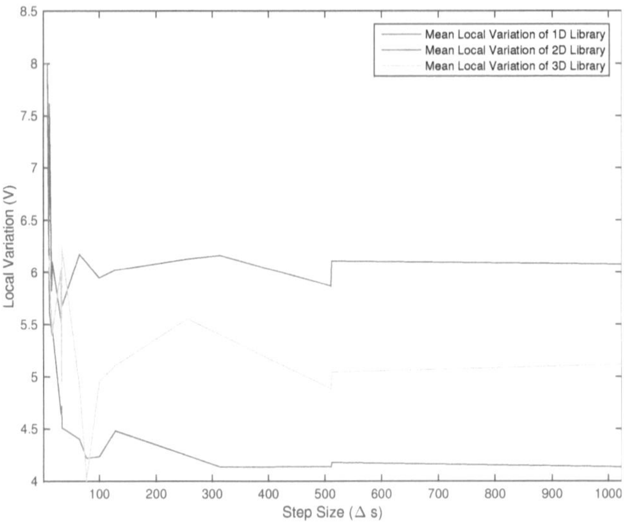

Figure 4.12: Average Second-Order Local Variation vs. Step Size for Each Wavelet Library Showing that the 2D Library is Smoother than the 3D Library.

Inter-Dimensional Library Creation

As shown in Figures 4.9 and 4.11, some two and three dimensional libraries do not share similar wavelets. The effect of this is that possible RFI sources may not be represented properly when performing the DWT analysis. For this reason a scheme to combine these libraries must be considered.

In order to achieve this, the one-dimensional and two-dimensional wavelet libraries were combined using a very similar method to that illustrated in Figure 4.5.

First, each of the smooth unique wavelet libraries was read, and their wavelets at each measure of step size were stored locally. As the smooth unique libraries ensured that all wavelets present in them had their second-order local variation below $V_H = 8$, the local variation was not tested.

58

The loaded wavelets were then ordered according to their second order local variation and were then compared using the similarity algorithm previously described. The result of this is shown in Figure 4.13.

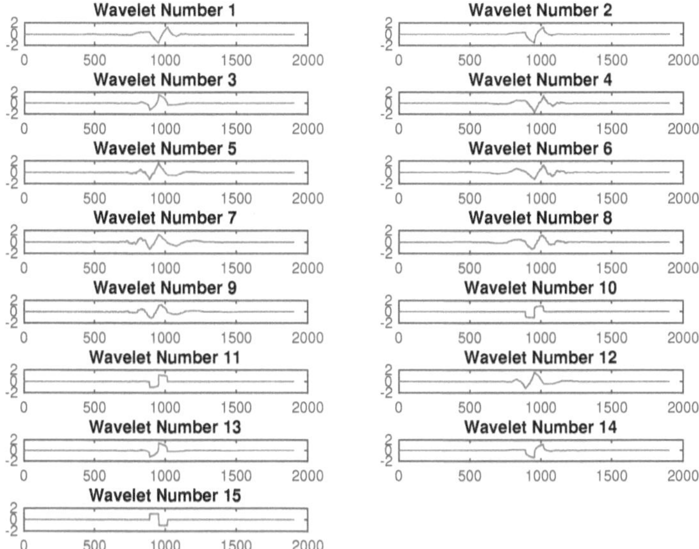

Figure 4.13: The Resultant Wavelets of Multi-Dimensional Libraries

It can be seen in wavelet numbers 7 to 9 that the wavelets present in this inter-dimensional library are not perfectly smooth, and have multiple zero-crossings. By comparing the results of Figures 4.9, 4.11 and 4.13 it can be seen that the wavelet present in the inter-dimensional set are more diverse and will be more effective for a general wavelet library for interference suppression.

Library Size

In order to highlight the computational complexity associated with the design of multidimensional wavelet libraries the resultant sizes and attributes of selected wavelet libraries are shown in Table II.

It can be seen that library sizes increase exponentially as the length-N wavelet filter sequences increase. The effect of this is that the computation cost of the second-order local variation and similarity between wavelets grows extraordinarily large.

For example when computing the wavelet library for the length-16 wavelet sequences with a search granularity of $\Delta s = \pi/1024$ the resultant library has approximately 1 billion entries. The calculation of second-order local variation takes an average 0.5ms on a system running MATLAB 2015a with the following specifications:

- Intel(R) Core(TM) i7-4790 CPU @ 3.60GHz

- 24 GB DDR3 RAM

- NVIDIA GTX 1060 (6 GB)

- Ubuntu 16.04 LTS

The effect of this is that it requires about 6 days of processing just to create a 3D library of smooth wavelets.

For this reason, higher dimensional libraries are somewhat inconceivable as the processing overhead to guarantee smooth unique wavelets is too large.

TABLE II

Size of Wavelet Libraries when Varying Search Granularity

Dimension	Step Size	Library Size	Smooth Wavelets	Unique Wavelets
1D	$10/\pi$	10	3	3
1D	$128/\pi$	128	31	9
1D	$512/\pi$	512	119	10
1D	$262144/\pi$	262144	59667	10
2D	$10/\pi$	100	17	6
2D	$128/\pi$	16384	1384	31
2D	$512/\pi$	262144	20937	33
2D	$4096/\pi$	16777216	1318418	33
3D	$10/\pi$	1000	4	3
3D	$128/\pi$	2097152	512	13
3D	$512/\pi$	134217728	32461	16

Chapter 5

Suppression Scheme Design

This chapter considers the literature, theory and designs of the previous sections and attempts to create a framework for a generalized interference suppression system. This is achieved by first considering approaches of matching the designing wavelets to particular interference sources. Then by measuring the performance of suppression by considering different thresholding methods and various tuning parameters of the DWT. The chapter is concluded with a discrete wavelet based interference mitigation system capable of automated RFI excision.

5.1 Background to Design

The validation of the library of designed wavelets and their integration into an interference suppression framework is discussed in this section. Data is synthesized in order to create a reliable way to compare the performance of the designed wavelets with wavelets predefined by the Matlab Wavelet Toolbox™ [8]. It will be seen that the choice of wavelet has severe repercussions on both the excision of the interfere as well as on the resulting shape of the reconstructed signal. Therefore, this section determines the optimal parameters to be used for the DWT. These being, the type of threshold, the decomposition parameters and the performance metrics used to measure the suitability of the suppression algorithm.

5.1.1 Synthesized Data

In the best case scenario, optimal interference suppression is achieved
when a single wavelet in the designed library of smooth wavelets is highly
correlated or matched to a particular interference source. The effect of this
is that the most compact representation of a function will be achieved if it
matches the wavelet perfectly at a particular scale. This is because, when
the wavelet is shifted and dilated in time, it should at some point perfectly
align with the function being analyzed given that they are precisely the same.

If the most compact representation in the wavelet domain is achieved,
then the interference's energy should be condensed into a single scale at
a particular sample in the wavelet domain. The effect of this, is that if
the threshold is determined correctly, then only the interference should
be reconstructed. Thereby enabling the reconstructed interference to be
coherently subtracted from the original waveform, thus effectively performing
interference suppression.

For this reason, a signal was created that comprised of two scales of
the Daubechies 10 (Db10) wavelet. This signal with -30dB of additive
Gaussian white noise may be seen in Figure 5.1b and the corresponding
wavelet can be seen in Figure 5.1a.

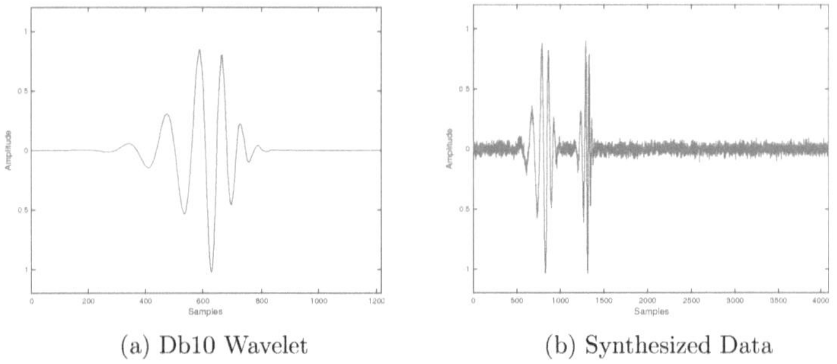

(a) Db10 Wavelet   (b) Synthesized Data

Figure 5.1: Figures Showing the Formulation of the Synthesized Data

5.1.2 Real-World Data

In order to validate the results obtained with the synthesized data, readings
from the SKA RFI monitoring station were used. This data was collected
using surveillance antennas and the raw voltage readings were written to file.
The readings of interest comprised of aeronautical telemetry signals, namely
Distance Measuring Equipment (DME) and Secondary Surveillance Radar
(SSR) signals. As well as arc welding transients, captured as a result of
maintenance on site. These waveforms may be seen in Figure 5.2 in both the
time and frequency domains.

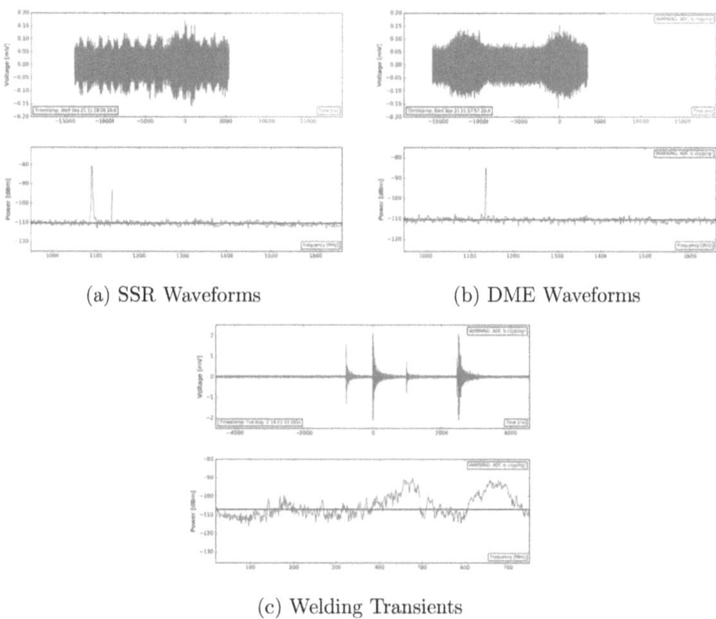

(a) SSR Waveforms   (b) DME Waveforms

(c) Welding Transients

Figure 5.2: Figures Showing the Types of RFI Collected at the RFI
Monitoring Station at the SKA

It can be noted that the SSR and DME waveforms shown in Figures 5.2a
and 5.2b are comprised of sinusoids as clearly shown by the impulses in
their frequency domain responses. Furthermore, it can be seen that the SSR
waveform contains two sinusoidal components, the impulse at 1.15 GHz in

63

the frequency domain can be attributed to that of the DME pulse.

In addition to this, it can be seen in Figure 5.2c that the arc welding transients are broadband signals that have frequency components that occupy all of the frequency spectrum. The effect of this, is that Fourier techniques would be unsuitable for excision of these types of transient RFI.

Lastly, it can be noted that the distinctions made in previous sections regarding stationary and non-stationary signals are clearly reflected by this data. It can be said that the arc welding transients are non-stationary interference, as they appear for brief moments in time and their periodicity is unpredictable as the interference is a result of human actions. Whereas, the DME and SSR waveforms are stationary interference, as they clearly more predictable for the measurements in this data set.

5.1.3 Performance Metrics

In order to validate the designed wavelet library and the selected data sets, it is necessary to define a testing procedure. The way in which this is done, is by first ensuring the wavelet coefficient shrinkage algorithm is effective on the synthesized data and then on the real-world recordings. In order to do so, it is necessary to determine techniques of measuring the performance of each wavelet, such that the optimal wavelet basis can be selected.

In the case of the synthesized data, no wavelet selection is needed, as the synthesized data is comprised of the existing wavelet as shown in Figure 5.1. For this reason, the synthesized data is only used for the selection of the optimal thresholding technique. Whereas for the real-world data, and the application shown in the next chapter, it is necessary to determine which wavelet is most suitable for the particular interference source.

Root-Mean-Square Error

The Root-Mean-Square Error (RMSE) is a commonly used statistical measure of the differences between an observed signal and a predicted value of an estimation process. As per the name, the RMSE is the square-root of the mean of the square difference between the estimated values and observed values, as given by

$$\text{RMSE}(x(n)) = \sqrt{\frac{\sum_{n=1}^{N}(x_p(n) - x_o(n))^2}{N}}. \tag{5.1}$$

Where $x_p(n)$ is the function consisting predicted values and $x_o(n)$ is the observed signal both of length N. It can be noted that the RMSE is always non-negative and the optimal value that would show no difference between the estimated signal and the observed signal is 0.

Furthermore, the RMSE measure is significantly more sensitive to large errors as the RMSE is proportional to the square error [52]. This is a suitable metric for measuring the performance of interference suppression in radio astronomy interference should have significantly higher power than the observed sources [53].

Normality Testing

As discussed in Chapter 3, normality testing is effective in characterizing whether interference is present in Gaussian data. For this reason the Lilliefors test, a commonly used normality test, is used to validate the performance of the wavelet based interference suppression system.

Normality tests are used to determine whether a data set is highly correlated to a normal distribution. This is achieved by formulating the null-hypothesis, which in this case is that the input data is normally distributed, and testing it through the use of the Lilliefors test. In the case that the data is normally distributed, then the null hypothesis is accepted whereas, if it is not normally distributed then the null hypothesis is rejected.

The Lilliefors test is formulated by first considering the z-scores of a particular distribution, given by

$$Z_i = \frac{X_i - \mu_X}{\sigma_X} \tag{5.2}$$

where μ_X is the mean of data to be tested, X, and σ_X is it's standard deviation. From this, the Lilliefors test computes the proportion of the Z_i-scores that are smaller or equal to its expected value, which is called the Cumulative Distribution Function (CDF) and referred to in the Lilliefors test as $\mathcal{L}(Z_i)$. For each of these scores the probability associated with it comes from a standard-normal distribution with a mean of 0 and a standard deviation of 1. This probability is given by

$$\mathcal{N}(Z_i) = \int_{-\infty}^{Z_i} \frac{1}{\sqrt{2 * \pi}} e^{-\frac{1}{2} Z_i^2}. \tag{5.3}$$

From this, the Lilliefors test criterion is formulated by

$$L = \max_i \{|\mathcal{L}(Z_i) - \mathcal{N}(Z_i)|\}. \tag{5.4}$$

The effect of this is that the maximum absolute error between the normally distributed Probability Density Function (PDF) and the CDF is measured. The null hypothesis is tested by comparing the L value to the pre-computed values using Monte Carlo methods at some significance interval [54]. The critical values for several confidence intervals can be seen in Appendix B.

5.2   Wavelet Thresholding

Considering that a library of wavelets has been designed for the decomposition and data has been selected for validation, it is necessary to determine the optimal thresholding technique to be used. The choice of a threshold is fundamental to the design of an interference mitigation system.

As the effect of a large threshold is that not all coefficients that relate to the interference are included in the resynthesis, whereas if a threshold is too low, then too much noise is included into the resynthesized interference. This section considers a few different thresholds based of Donoho and Johnstone's research and attempts to select the most suitable one for interference mitigation.

For the documented threshold selection procedure, all thresholds were calculated using the built in tools in the Matlab Wavelet Toolbox™ [8].

5.2.1   Adaptive Universal Threshold

It was discovered that a simple extension to the universal threshold, was to calculate the universal threshold on a level dependent basis. This meant that rather than just using the highest resolution detail coefficients $d_{J-1,k}$ for the calculation of the MAD, it was possible to calculate the MAD for each scale. Thereby determining the threshold for each level of the decomposition.

5.2.2 Threshold Selection

Trials were run to determine the most suitable threshold for interference suppression. These experiments were conducted by varying the additive Gaussian white noise from 0dB to -50dB of the synthesized data and calculating the RMSE of the resultant thresholded signal.

In order to ensure the statistical significance of the results, the calculated RMSE of each threshold at each noise level was the average of 100 iterations. The optimal decomposition depth was determined by adding -5dB of Gaussian white noise to the synthesized data, and calculating the RMSE at each decomposition level for each threshold. For each decomposition the Db10 wavelet was used and it can be seen in Figure 5.3a, that the universal adaptive threshold applied using soft thresholding techniques with a decomposition depth of 12 resulted in the lowest RMSE.

In the next trial, the noise level was varied and the RMSE of each threshold applied using soft and hard thresholding were graphed. The results of this process may be seen in Figure 5.3b.

It can be seen that the RMSE recovered signal associated with each threshold gets linearly worse as the noise increases logarithmically in steps of 5dB. Furthermore, it is clear that the universal adaptive threshold applied using soft thresholding consistently outperforms the other thresholds. For this reason, the universal adaptive threshold is used in further uses of the DWT interference suppression scheme.

Finally, in order to validate the chosen parameters were effective in maintaining the -5dB of additive noise, the Lilliefors test with a 0.01% significance level was run and a histogram was plotted.

It can seen that the output histogram clearly follows a normal distribution as shown by Figure 5.4. Where the blue lines represent the bin value for each of the samples of the recovered Gaussian white noise, and the red line represents its normal distribution fit. In addition to this the concatenated output vector from each of the 100 trials all pass the Lilliefors test with a

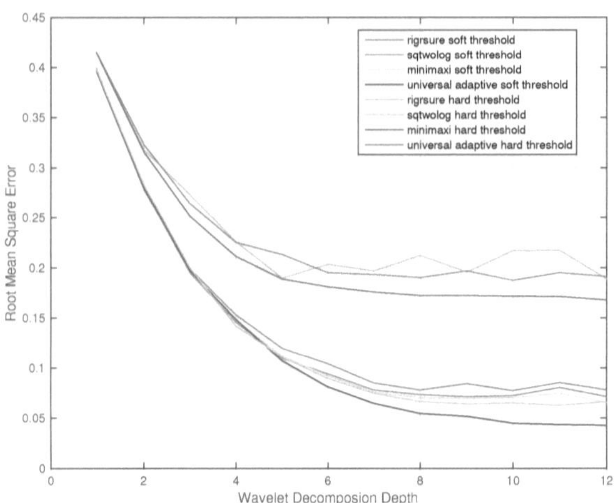

(a) Effect of Varying Decomposition Depth and Threshold Type

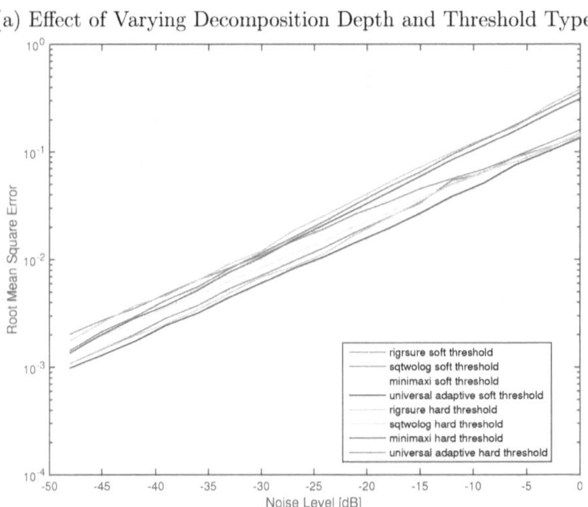

(b) Effect of Varying the Noise Level and Threshold Type

Figure 5.3: Plots Showing the Relative Performance Between Threshold Type, Decomposition Depth and RMSE

0.01% significance level.

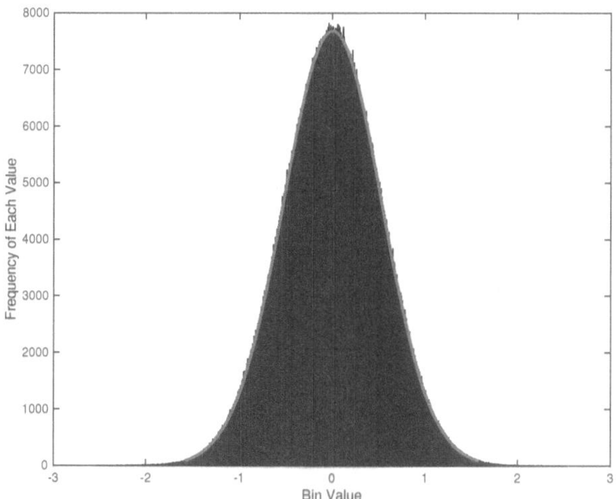

Figure 5.4: Histogram of the Result Gaussian Noise From the Synthesized
Data Validation Test

5.3 Interference Preprocessing

In order to accommodate testing for the real-world signals collected at the
RFI monitoring station at the SKA, it is necessary to perform modeling of
each of the RFI waveforms. This was done to measure the RMSE of the
recorded signals after the suppression of the additive Gaussian white noise.
For this reason, a preprocessing scheme is documented as well as a median
based modeling strategy is considered.

5.3.1 Automated RFI Extraction

In work done by Czech et al. [26], it was shown that by using an approach
of averaging and applying multiple thresholds it is possible to automatically
extract transient waveforms. This technique was used as it is impractical to
extract large numbers of interference waveforms manually. This interference

69

extraction algorithm may be considered in two separate stages. The first stage coarsely extracts the large features of the inputted signals, whereas the second stage refines the selection by extracting the finer details of the RFI present in the inputted signal. A visualization of this algorithm may be seen in Figure 5.5.

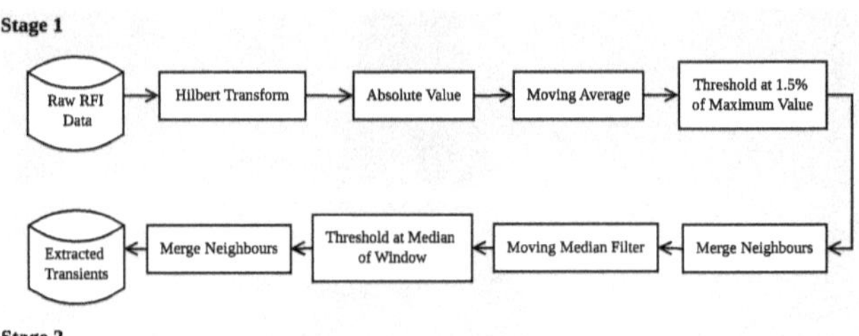

Figure 5.5: Block Diagram showing the Formalization of the Automated RFI Extraction Algorithm

For purposes of illustration of the algorithm, the arc welding transient signals shown in Figure 5.2c were used. The interference extraction algorithm begins with applying the Hilbert Transform to the time domain data and taking the absolute value of the transformed data. This is done in order to extract the envelope of the data. A moving average filter is then applied with a window size, L_1, of 640 samples, which is determined empirically by Czech et al. [26]. This is done to smooth the transitions of the envelopes of the transients, as can be seen by the orange line in Figure 5.6.

In the case of the arc welding transients, the course threshold, T_m, is set to 1.5% of the maximum value of the original signal, which may be varied for different interference types. The Hilbert transformed averaged signal is then thresholded such all samples above the course threshold are considered as regions of interest. The course thresholding mechanism minimizes the number of transients missed in the finer pass. Then, all regions of interest that are T_m samples or less away from each other are merged and regions consisting of a number of samples less than T_m are discarded. The selected course thresholded regions of the interest may be seen in Figure 5.6 as the yellow line.

In the second stage of the algorithm, a fine threshold is used in order to extract the arc welding transients from the course regions that have been found. This is achieved by determining the adaptive threshold T_2 that

70

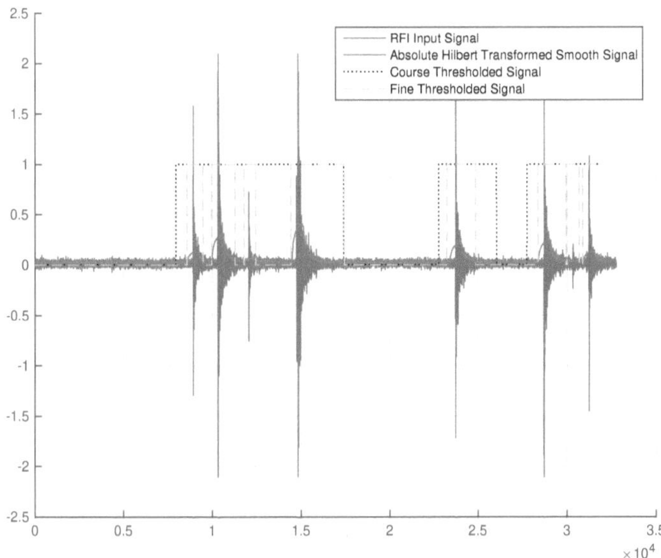

Figure 5.6: Plot Showing the Time Domain Signal and its Thresholds

allows for the separation of transients that are in near proximity to each other. The adaptive threshold is calculated by finding the median value of the Hilbert transformed averaged signal in a window T_{md} of 7000 samples. According to Czech et al. [26] the exact value of this window is not critical, but has to be larger than the moving average window, L_1.

Then each Hilbert transformed averaged sample in the window is used to calculate the adaptive threshold T_2. This is then used to threshold the original signal such that all values above the threshold are considered regions with interference. Finally, all fine thresholded regions that are $T_{\mathrm{med}} = 28$ samples away from each other are merged, and those less than that are discarded. The fine thresholded regions may be seen as the purple lines in Figure 5.6.

71

5.3.2   Median Based Modeling

In order to create a model of the extracted interference from the algorithm described above, a median filtering method was used. This method aligns the extracted interference waveforms and determines which interference waveforms are similar by using a measure of their normalized inner products. If the extracted interference has an inner product greater then 0.25 then the waveforms are considered of the same class, and are used in determining the median based model.

An overlay of each of the extracted arc welding transient waveforms and their Power Spectral Density (PSD) can be seen in Figure 5.7a. In addition to this, the median based model of the arc welding transients can be seen in Figure 5.7b.

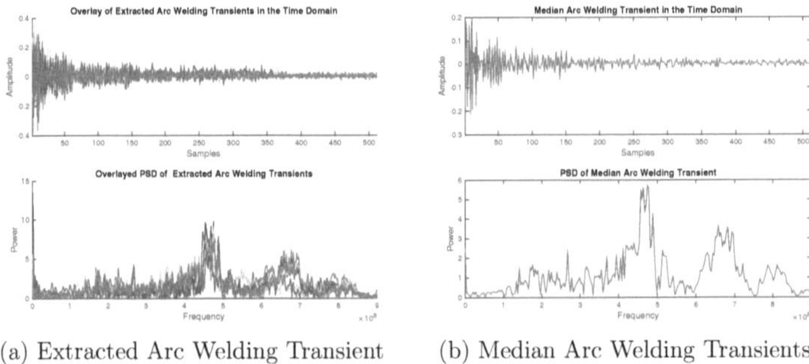

(a) Extracted Arc Welding Transient   (b) Median Arc Welding Transients

Figure 5.7: Plots Showing Time and Frequency Domain Characteristics of the Arc Welding Interference

It can be seen in Figure 5.7b, that there still is slight noise in the model. As RMSE is the primary measure of performance of the interference mitigation algorithm, the measure may have difficulty in determining which noise is a result of the additive process and which is inherently present in the model. However it will be seen in later results, that this effect is immaterial as the overarching shape of the interference waveform has a greater effect on the results, than slight deviations in the noise level.

5.4 Wavelet Selection

The final consideration to be made for the design of a interference suppression scheme is the selection of the optimal wavelet. As already mentioned the optimal wavelet maximizes performance of the suppression scheme by enabling the most compact representation of a signal in the wavelet domain. Such that the threshold estimator is the most accurate in determining the correct level to shrink the detail coefficients to. For this reason, this section documents the procedure of optimal wavelet selection.

5.4.1 Entropy Based Selection

Originally entropy based measures were used for the determination of the optimal decomposition level of the WPT [21]. However it has been shown that it is also an effective measure of the suitability of a wavelet for matching a particular signal [36, 39]. This is because entropy enables the ability to distinguish between coefficients in the expansion that are condensed into a single location which correspond to a low entropy. Whereas a decomposition with high entropy is a result of the *matches* to the wavelet being spread across each level of the expansion.

This being said, the entropy measure as defined by Coifman and Wickerhauser is given by [21]

$$H_{\log_2} = -\sum_{n=1}^{N} \frac{|p(n)|^2}{||\mathbf{p}||^2} \log_2 \left(\frac{|p(n)|^2}{||\mathbf{p}||^2} \right) \tag{5.5}$$

where $p(n)$ classically represented the wavelet packet coefficients and $||\mathbf{p}||^2$ is the norm of the coefficients squared. As the DWT implementation used in this research product is not the WPT, we can consider $p(n)$ to be a vector consisting of the concatenated detail coefficients from each scale. Furthermore, the measure of normalized entropy as specified by Berger et al. is given by [55]

$$H_{\mathrm{norm}} = \frac{2^{H_{\log_2}}}{N} \tag{5.6}$$

where N represents the length of the concatenated detail coefficients vector.

5.4.2 Performance Relative to Conventional Wavelets

In order to prove the validity of the entropy based selection procedure two different experiments were run. For each of the experiments, the decomposition depth of the DWT was fixed to 12, the threshold that was used was the level adaptive universal threshold and it was applied using the soft thresholding mechanism. For the first experiment, the synthesized data set shown in Figure 5.1 was tested using varying noise power levels and the resulting entropy of the decomposition was measured when decomposed with several different wavelets. As the synthesized data

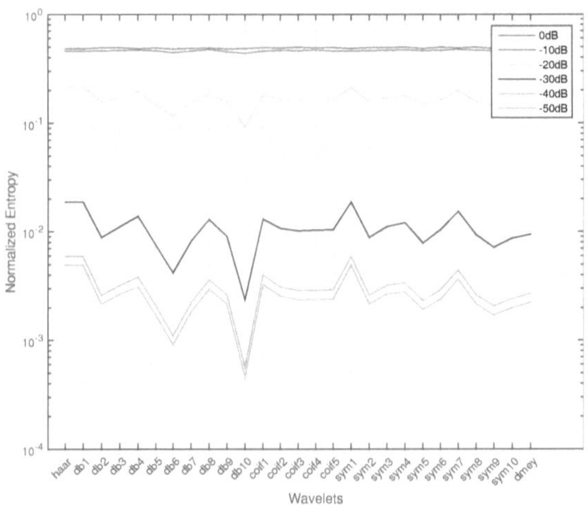

Figure 5.8: Plot Showing the Performance of Entropy Based Selection at Varying Noise Levels for the Synthesized Data

consisted of two Db10 wavelets at different scales, it was expected that the entropy based measurement should clearly show a strong match to the Db10 basis for the decomposition. It can be seen in Figure 5.8, that this is the case.

It can be noted that at noise levels greater than -20dB the selection of suitable wavelets become challenging. This is because, when additive Gaussian white noise levels get too high they corrupt the signal of interest to an unrecognizable state. However, it can be noted that for the radio astronomy based application of this interference mitigation scheme RFI power levels should always exceed the astronomical observed noise. For this reason, the robustness of the entropy based selection procedure is deemed acceptable.

In the next experiment, the median-modeled arc welding transient was used to confirm the robustness of the RMSE measure of performance relative to that of the entropy based wavelet selection strategy. This was achieved by first using the model to select the optimal wavelet from the multi-dimensional library that was previously created. After which, the entropy and RMSE of the model was calculated when -5dB Gaussian white noise was added to it. This was done for several different wavelets as well as the wavelet chosen with the lowest entropy from the multi-dimensional library. The results of these experiments may be seen Figure 5.9.

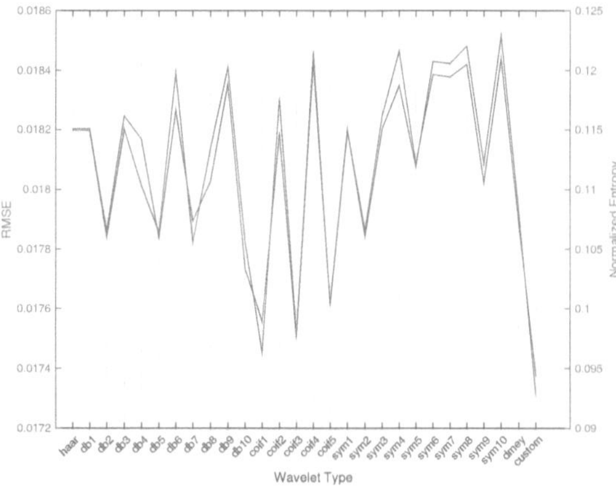

Figure 5.9: Plot the Relationship between Entropy and RMSE for various Wavelets for the Arc Welding Transient Data

It can be seen that the RMSE and entropy of the resultant expansions are highly correlated. Furthermore, it can be seen that the lowest entropy custom wavelet selected from the multidimensional library posses the optimal performance in terms of RMSE.

Therefore, it can be concluded that the multi-dimensional library is suitable for the matching of wavelets to transient RFI signals similar to the arc welding transients. In addition to this, it can be said that the measures of RMSE and entropy are suitable performance and selection metrics respectively.

5.5   System Formulation

Most applications of the interference mitigation system require the system
to be capable of removing various types of interference. For this reason it is
necessary to design a system, using the methods previously explained, that
is capable of optimal suppression of various types of interference. In this
section, the construction of a generalized interference suppression algorithm
is described.

5.5.1   Matched Filter Design

Optimal performance of the interference mitigation algorithm is achieved
when a well suited wavelet is chosen for the wavelet decomposition. Because
there exist many different interference sources and waveforms that may be
present in the same recording, it is imperative to know which wavelet to use.

For this reason, matched filters are used for the detection of different
types of interference.   In addition to the justifications for interference
modeling given previously, the median based model can be used for
interference detection. This is achieved by using the median based models
as the transfer functions of the matched filters.

A matched filter is a mechanism that allows for the detection of signal of
known shape even when effected by noise. This is achieved by designing a
filter $h(t)$ that maximizes its output Signal-to-Noise Ratio (SNR) when the
input to the filter is a particular signal of interest. It is possible to show
that maximal SNR is achieved when the input to the system is the signal
of interest. This is achieved by defining the matched filter in the frequency
domain as

$$H(\omega) = kX^*(\omega)e^{-j2\pi f t_m} \tag{5.7}$$

such that $X^*(\omega)$ is the conjugate of the signal of interest that we would like
to detect that has been time shifted to some time t_m and scaled by some
parameter k [56].

5.5.2   System Integration

From the previously designed and validated subsystems described in previous
sections, it is possible to integrate each of the designs into a wavelet based

interference mitigation system. A high-level overview of the integrated
implementation of the interference suppression system can be seen in
Figure 5.10.

The first step of the algorithm is a preprocessing stage for the inputted
data. This is done in order make the interference mitigation system able
to process various data structures and formats. This preprocessing stage
facilitates both the automated interference extraction and the median-based
modeling stages.

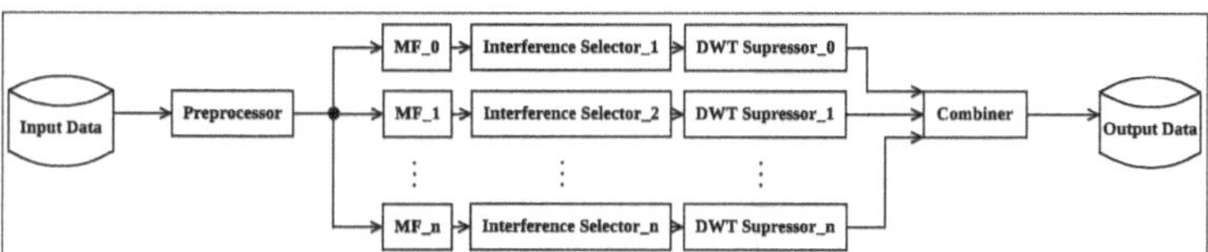

Figure 5.10: High-Level Block Diagram of the Implementation of the
Wavelet Based Suppression System

Once the median-based models of the interference have been determined,
they are used as the transfer functions of the matched filters. In addition
to this, each of the matched filters have a wavelet associated with them.
Where each of the wavelets are matched to the median-based models using
the optimal entropy methods.

The effect of this configuration, is that when the matched filters detect a
particular interference source, the optimal universal adaptive thresholding
scheme is applied using the entropy matched wavelet. When multiple
matched filters detect interference, the filter with maximal output is
considered to be the correct match.

Furthermore, as the matched filtering operations give indications of
the time localizations of the detections, it is possible to only select regions
that contain interference for suppression. With the combination of the
course thresholding techniques described in Section 5.3, it is possible to
select regions that contain RFI that are in close proximity to each other.

In the final stage of the algorithm, the interference suppressed regions
are reinserted into their original positions of the inputted signal. These
indexing and insertion operations are performed by the *combiner* stage.

In order to validate the effectiveness of the system, the interference signals shown in Figure 5.2 were concatenated together and injected into the wavelet based interference suppression implementation. The resultant concatenated waveform may be seen in Figure 5.11. It can be noted the that the order of concatenation begins with the arc welding transients, followed by the DME pulses and the SSR pulses are concatenated last.

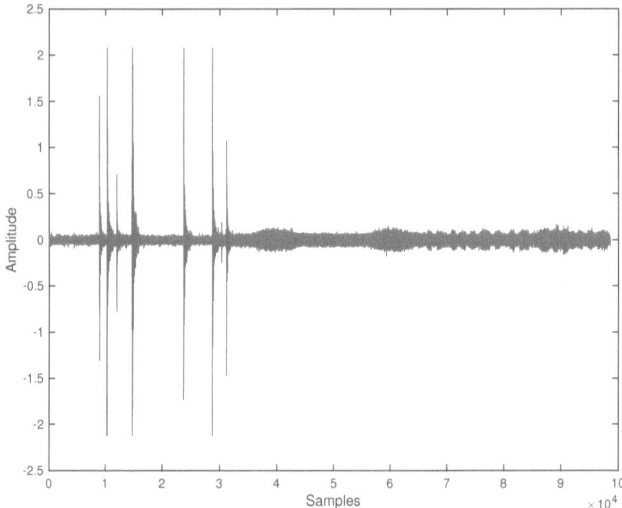

Figure 5.11: Concatenated Interference From the SKA RFI Monitoring Station

The median based models of each of the interference waveforms were used to determine the optimal-entropy wavelet. Each median based model, and their corresponding wavelet can be seen in Appendix C.1.

It can be noted that the arc welding transient and the DME pulse have very similar wavelets associated with them. This is attributable to the fact that both waveforms amplitude envelopes are similar, where the arc welding transient model has a repeating sinusoidal envelope and the DME pulse has a single envelope of a similar shape. The effect of this is that when the wavelet decomposition is applied, there are strong matches at a particular scale with highly condensed energy, which results in a low entropy for the decomposition.

78

The outputs of the matched filters that use median-based models for their transfer functions can be seen in Figure 5.12.

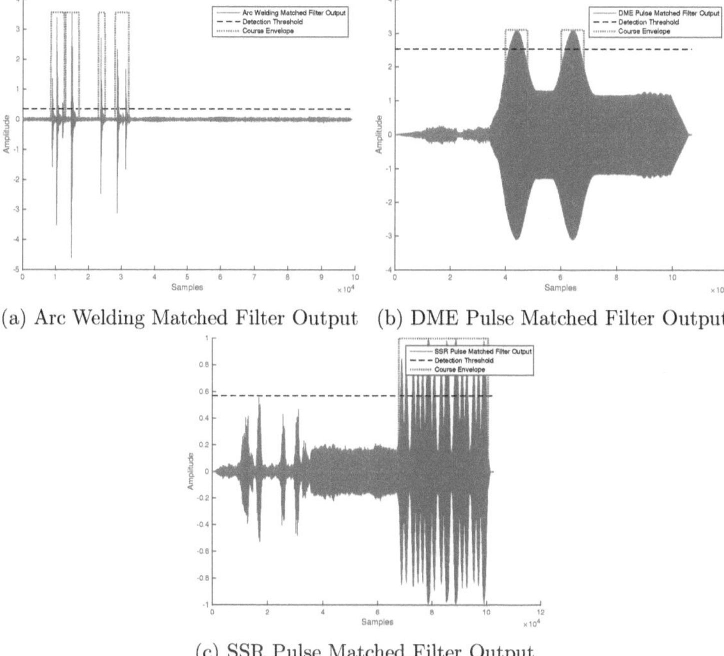

(a) Arc Welding Matched Filter Output   (b) DME Pulse Matched Filter Output

(c) SSR Pulse Matched Filter Output

Figure 5.12: Plots Showing the Outputs of the Matched Filters

It can be seen that each of the matched filters have maximum output when they detect their corresponding interference waveforms. The orange lines in the figures represent the thresholds determined by the standard deviation of outputs of the matched filters. In the interference selection stage of the algorithm all outputs of each matched filter that exceed the standard deviation are considered as detections of interference. The yellow lines shown in each of the plots in Figure 5.12 are the envelope associated with the course thresholding strategy.

Once the regions that contain interference have been determined, the

optimal wavelet DWT thresholding scheme is applied. The result of the
suppression and reinsertion can be seen in Figure 5.13.

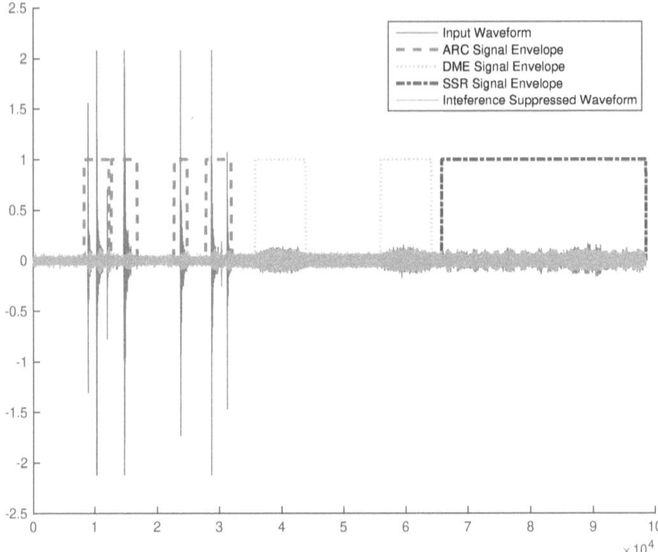

Figure 5.13: Plot Showing the Relationship between the Input and Output
Waveforms

This figure, shows the relationship between the input and output waveforms
by superimposing them on top of one and other. Where the blue waveform
in this figure represents the original inputted signal and the green waveform
represents the interference suppressed waveform. In addition to this, the
regions of interest that have been selected by the matched filter-bank can
be seen by the orange, yellow and purple lines respectively.

In this simulation, the number of decompositions in the DWT was set
to 12, and soft thresholding was used for the coefficient shrinkage. It can
be seen that there is a clear reduction of interference when comparing the
input to the output waveform.

However, it can be noted that there are slight discontinuities in the
regions between the thresholded waveforms and the regions of which no
interference was detected. This effect may be attributable to fact that the

thresholding mechanism slightly reduces wavelet coefficients that do not contain interference. This being said, it will be seen the next chapter that the precise loss of information is approximately negligible.

From the design and validation of the interference suppression system, it is possible to proceed to a radio astronomy based application.

Chapter 6

Radio Astronomy Application

This chapter investigates the application of the designed wavelet based interference suppression system to radio astronomy data. The chapter begins with a brief description of the KAT-7 array and its role in the SKA project followed by a summary of the theory surrounding the reception of radio emission from celestial bodies using interferometry.

In addition to this, the procedure of storing and accessing data collected by KAT-7 is explained as well as the Measurement Set format. From this, the various representations of the observations of the Messier 83 galaxy are shown as well as the effects of flagging this data set.

The chapter is then concluded with a final design of a interference suppression system for radio astronomy. This system is used to draw results and conclusions in the closing sections of this report.

6.1   Background to Radio Astronomy

Radio astronomy is the field of study concerning the analysis of radio-frequency radiation originating from celestial bodies. In this field, measurements of stars, galaxies and other astronomical sources are measured by their radio frequency emissions. Radio frequency radiation comes mostly from non-thermal radiation due to synchrotron radiation or electronic transition. The effect of this is that by using antennas, which are also known as radio-telescopes, it is possible to measure the faint radio emissions of celestial bodies [53].

The typical signal strength received by radio telescopes from astronomical sources typically measure between -150 dBWm^{-2} and -220 dBWm^{-2}. Where the strongest radio source at 300 MHz has a PSD of about 105 Jansky (1 Jansky is 10^{-26}Wm^{-2}Hz^{-1}). The effect of this is that radio telescope receiver systems have to be significantly more sensitive than other electromagnetic receivers. For this reason, radio telescopes are highly susceptible to RFI and are a well suited application to the wavelet based interference suppression system [28].

In radio astronomy it is possible to represent the received signals in many different domains, where each representation highlights different features of the received signals. For this reason, it is necessary to gain insight into radio astronomy based techniques in order to effectively apply the interference suppression algorithm. Therefore, this chapter documents radio astronomy theory and shows the suitability of wavelet based interference mitigation to radio astronomy.

6.2   The Karoo Array Telescope

The seven-dish Karoo Array Telescope (KAT-7) [1] is a radio telescope array in the Northern Province of South Africa. The array was constructed as a precursor to the larger MeerKAT and the Square Kilometer Array (SKA) in order to prototype the technology and requirements of the larger arrays.

Figure 6.1: Aerial Photograph Showing the Configuration of Radio Telescope Array used In the KAT-7 Array [57]

The geographic location of the array was selected due to the lack of terrestrial infrastructure in proximity to the site, which reduces the issue of RFI. In addition to this, a radio-quiet zone was declared in order to ensure maximal sensitivity of the array for future use. The physical arrangement of the seven-dish array was determined to posses optimal collecting time and geometries and can be seen in Figure 6.1 [1].

In order to improve sensitivity of the receivers, the radio front-end of the receivers are cryogenically cooled to 70° Kelvin. The precise specifications of the array may been seen in Table III.

In the remaining chapters of this report, data collected from the KAT-7 array is used.

TABLE III

THE KAT-7 SYSTEM SPECIFICATIONS [1]

Parameter	Value
Number of Antennas	7
Dish Diameter	12m
Baselines	26 m to 185 m
Frequency Range	1200 MHz to 1950 MHz
Instantaneous Bandwidth	256 MHz
Polarization	Linear
System Temperature	$\leq$35K
Full Width Half Maximum (FWHM) of primary beam	$\approx 1°$

6.3 Radio Astronomy Theory

In order to gain intuition about the signals that are processed by the interference suppression algorithm, it is necessary to investigate the origin and composition of radio signals from space. This section covers the mathematical formulation of these electromagnetic signals, the various techniques used to represent astronomical sources as well the interpretation received signals from the KAT-7 radio telescope array.

Much of the theory covered in this section can be found in literature by Kraus [58], unless otherwise specified.

6.3.1 Origins of Signals

Planck's radiation law states that all objects above a temperature of absolute
zero radiate energy in the form of electromagnetic waves. The law shows that
the brightness of an ideal radiator, also known as a black body radiator, at
temperature, T, and frequency ν is expressed by

$$B_\nu(\nu, T) = \frac{2h\nu^3}{c^2}\frac{1}{e^{\frac{h\nu}{k_\mathrm{B}T}} - 1} \tag{6.1}$$

where

> $\mathbf{B}_\nu$ is the source's brightness measured in $\mathrm{m}^{-2}\mathrm{Hz}^{-1}\mathrm{rad}^{-2}$
>
> $\mathbf{h}$ is Planck's constant
>
> $\boldsymbol{\nu}$ is the electromagnetic frequency
>
> $\mathbf{c}$ is the speed of light
>
> $\mathbf{k}$ is Boltzmann's constant
>
> $\mathbf{T}$ is temperature in Kelvin

From this theorem, the reception of the radiation from celestial objects can
be measured. If one considers a flat horizontal plane with area A on the
surface of the earth that is receiving electromagnetic radiation from the sky
above it, then it can be shown that

$$dW = B\cos(\theta)d\Omega d\nu dA \tag{6.2}$$

where

> $\mathbf{dW}$ is the infinitesimal received by the plane
>
> $\mathbf{B}$ is the brightness of the sky at position $d\Omega$
>
> $\mathbf{d\Omega}$ is the infinitesimal solid angle of the sky
>
> $\boldsymbol{\theta}$ is the angle between $d\Omega$ and the zenith
>
> $\mathbf{dA}$ is the infinitesimal area of the surface
>
> $\mathbf{d\nu}$ is the infinitesimal bandwidth of the electromagnetic spectrum

This formulation enables one to obtain an expression for the received power
of the entire plane by integrating over the bandwidth and the area of the
plane, such that

$$W = A \int_\nu^{\nu+\Delta\nu} \iint_\Omega B\cos(\theta)d\Omega d\nu. \tag{6.3}$$

If the simplified model of the receiver is replaced with an antenna, the expression for the received power must be reformulated. As the sky brightness varies with the direction that the antenna is pointing in, the brightness distribution of the sky, $B(\theta, \rho)$, in terms of a spherical coordinate system is used.

In addition to this, antennas have radiation patterns, $P_n(\theta, \rho)$, which are a measure of the response of an antenna to a radiation function of angles θ and ρ.

An illustration of the radiation pattern as well as the spherical coordinate system can be seen in Figure 6.2.

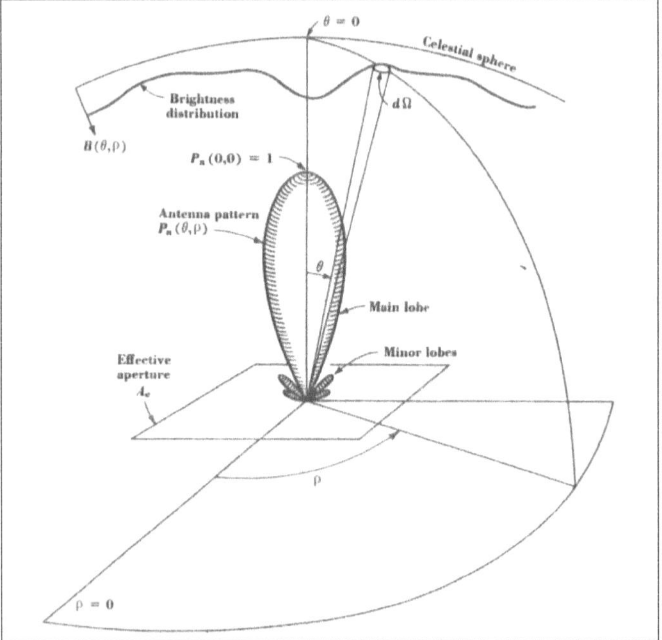

Figure 6.2: Figure Showing the Physical Coordinate System used For Radio Astronomy [58]

In order to generalize the received power of an antenna, it is necessary to describe the received power of an antenna using spherical coordinates by

$$w = \frac{A_e}{2} \iint_{\Omega} B(\theta, \rho) P_n(\theta, \rho) d\Omega. \tag{6.4}$$

Where A_e is the effective area of the antenna and the factor of $1/2$ is a result of a single polarization that is received by the antenna. This equation shows that in order to increase sensitivity of a receiver, it is necessary to change the radiation pattern or the increase the effective area of the antenna.

6.3.2 Electromagnetic Wave Polarization

Polarization of electromagnetic waves is an important phenomenon in radio astronomy as all radio sources are polarized. The effect of this is that the polarization of sources give astronomers insight into the magnetic fields and geometries of celestial objects. The way polarization is mathematically described is by first considering an electromagnetic wave moving in some direction $\vec{r}$ such that its electric field may be described by

$$\vec{E}(\vec{r}, t) = E_o e^{j(2\pi\vec{r}/\lambda - 2\pi\nu t)}. \tag{6.5}$$

This equation may be separated into its vertical and horizontal components by

$$\begin{aligned}
E_x &= E_{ox} cos(2\pi\nu t + \delta_x) \\
E_y &= E_{oy} cos(2\pi\nu t + \delta_y)
\end{aligned} \tag{6.6}$$

where δ_x and δ_y represent the respective phase shifts for each polarization. Such that when $E_{ox} = 0$ the electromagnetic signal is polarized along the y-axis and when $E_{oy} = 0$ the electromagnetic signal is polarized along the x-axis [59].

In order to infer information about a particular source, astronomers also use measurements of the circular and linear polarizations of electromagnetic waves. Circular polarization is the state in which the electric field of a wave has constant magnitude but its direction rotates perpendicular to the plane of propagation. Whereas the linear polarizations are made up of the equations shown in Equation (6.6).

Astronomers often make use of stokes parameters to represent the polarizations of electromagnetic waves in a more convenient way. The Stokes parameters can be described by

$$
\begin{pmatrix} I \\ Q \\ U \\ V \end{pmatrix} = \begin{pmatrix} E_x^2 + E_y^2 \\ E_x^2 - E_y^2 \\ 2E_x E_y cos(\delta) \\ 2E_x E_y sin(\delta) \end{pmatrix}
\tag{6.7}
$$

The use of the Stokes vector allows astronomers to infer information about both the direction and type of polarization. It will be shown in later sections that RFI appears differently in each of the stokes parameters, thereby allowing more accurate descriptions of the effectiveness of the interference suppression system [59].

6.3.3   Phase Interferometry

In radio astronomy the telescopes which are used possess finite resolutions that are determined by the minimum separation between sources before they become indistinguishable from one and other. The resolution of an antenna is defined by

$$
\theta \approx \frac{\lambda}{D}
\tag{6.8}
$$

where D is the diameter of an antenna and λ is the wavelength of the radiation being observed. This relationship shows that the only way to increase the resolution of a telescope is by increasing D. The effect of this is that the construction and maintenance of high resolution radio telescopes is both economically and structurally impractical relative to the achievable resolution of optical systems.

A technique used to combat the limited resolution of antennas is by using interferometry techniques that rely on multiple smaller telescopes. It has been shown that the associated resolution of a telescopic array is expressed by

$$
\theta \approx \frac{\lambda}{\beta_{\mathrm{max}}}
\tag{6.9}
$$

where β_{max} is the maximum distance between two antennas and is referred to as a *baseline*.

In order to understand the how observations are done using multi-antenna systems it is necessary to understand the geometries and planes of observation. The geometric relationship between sources in the sky and a baseline pair on the surface of the earth can be seen in Figure 6.3. The (l, m) coordinate system is used to describe positions of celestial objects in the sky and the (u, v, w) coordinates are used to describe planes on the surface of the earth.

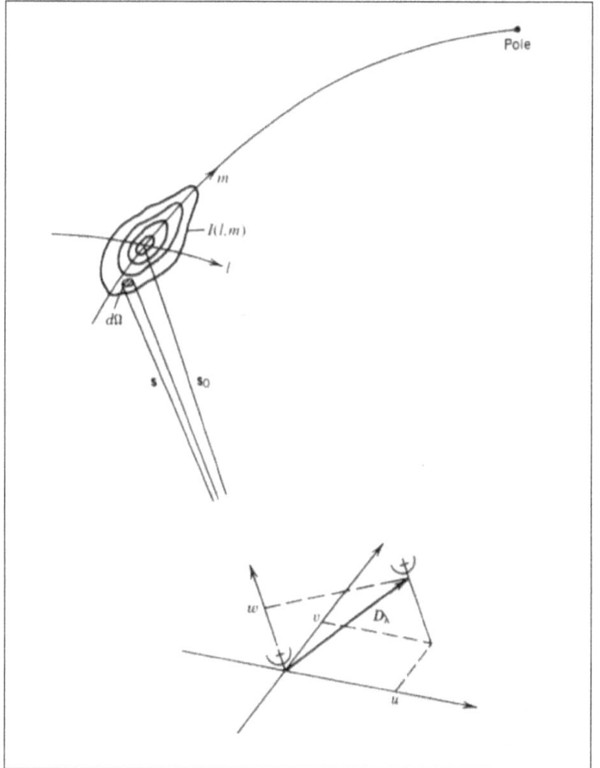

Figure 6.3: Figure Showing the Geometric Relationship Between a Source in the (l,m,) Plane and a Baseline Pair on The Surface of the Earth [53]

The (l, m) plane is used represent the intensity distribution $I(l, m)$, which is a projection of the celestial sphere onto a plane tangent to the source's center. It can be noted that the (u,v) plane is tangential to the observation direction S_0 such that it does rotate with the earth.

The baseline distance between a pair of antennas pointing at the same location in the sky, results in each antenna receiving time-delayed replicas of the same signal. The effect of this is that the average of the correlation of the electric field of the two antennas is proportional to the electric field of an object in the far field of the baseline pair given that their radiation patterns are identical. This correlated quantity is known as the visibility and is represented by $V(u, v, w)$ [53].

The relationship between the (u, v) and (l, m) planes is resolved through the use of the van Cittert–Zernike theorem. It shows that the response of an ideal interferometer, $V(u, v)$, is given by the two dimensional Fourier Transform of the sky brightness $I(l, m)$ for a co-planar array and a small angle such that

$$V(u, v, w) \simeq V(u, v, 0) = \int_{-\infty}^{\infty} \int_{-\infty}^{\infty} \frac{I(l, m)}{\sqrt{1 - l^2 - m^2}} e^{-j2\pi(ul+vm)} dl\, dm. \quad (6.10)$$

Where the sky brightness, $I(l, m)$ forms an image of the region being observed, and it is the plane that astronomers mostly use for inferring information about celestial sources.

In order to obtain this image, it is necessary to take the inverse two dimensional Fourier Transform of the complex visibility. In practice, the visibilities are only observed for discrete locations in the (u, v) plane such that the resultant discretized grid of the visibilities can be described by [60]

$$S(u, v) = \sum_{k=1}^{M} \delta(u - u_k, v - v_k) \quad (6.11)$$

such that the resultant discretized image is

$$I(l, m) = \int_{-\infty}^{\infty} \int_{-\infty}^{\infty} S(u, v) V(u, v) e^{j2\pi(ul+vm)} du\, dv. \quad (6.12)$$

6.3.4 Deconvolution

In interferometry the discretized measured visibilities often form incomplete representations of the (u, v) plane. This causes unwanted effects when applying the two dimensional Fourier transform to obtain the sky brightness.

The response of the antennas that result in the incomplete representation of the (u, v) plane is referred as the dirty beam or the Point Spread Function (PSF). It can be viewed as the transfer function of the receiver system, that contains unwanted secondary responses known as side-lobes.

The mitigation of these effects is commonly known as deconvolution and there exist a number of techniques to remove these unwanted artefacts. A common algorithm used to perform the deconvolution is the CLEAN algorithm. It is an efficient way of obtaining a good approximation of actual image [61].

The CLEAN algorithm is performed by the following steps:

1. Determine the strength and position of the brightest pixel/source in the dirty image, $I(l_i, m_i)$ and model it as the impulse $\delta(l_i, m_i)$.

2. Multiply the brightest pixel by some gain $\gamma \leq 1$ and save its position and magnitude to be kept track of in later iterations.

3. Multiply the pixel value with the transfer function of the dirty beam pattern, and subtract the result from the original dirty image.

4. Repeat steps 1 to 3 until the remaining sources are below some level or the maximum number of iterations of the algorithm has been reached. The values for the level and maximum iteration length are determined depending on the image.

5. Convolve the resulting scaled dirac-delta's with an idealized model of the clean beam pattern.

6. Reform a cleaned version of the original image by rearranging the cleaned beams in the same locations that the pixels were lifted from.

The result of the CLEAN algorithm, is a smoothed version of the original image where all bright sources appear more clearly. In the implementation of the CLEAN algorithm in the Common Astronomy Software Application (CASA) [62], a commonly used radio astronomy software suite, the two most influential inputs to the algorithm are image size and cell size. Where the cell size determines pixel size of the image to be made and the image size determines the of extent of the field to be produced. These parameters are determined from the known specifications of the KAT-7 array.

6.4 KAT-7 Data Analysis

The KAT-7 array produces raw data in the form of the correlations of the electric fields of two antennas. In order for astronomers to effectively analyze the observations from the array, a data processing chain is used. This section documents the processing chain as well as analyzing a particular observation of interest.

6.4.1 Data Processing

The data processing chain required to create file formats readable by astronomers begins with the correlation of the received radiation from arrays of radio-telescopes. This is obtained by correlating each of the baseline-pair's sampled electric fields and correlating them on FPGA-based correlators. This correlated data is then channelized, a process that band-pass filters the correlated time domain data into multiple frequency bands. The FPGA systems then packetize the channelized correlated data and save the results to Hierarchical Data Form (HDF5) format files.

Before the HDF5 files are saved to a format useful for astronomy, reduction techniques are applied to remove fringe effects from the RF-chain as well as physical disturbances that may have occurred during the observation time. At this point, the modified HDF5 data is converted to a Measurement Set (MS), so that the meta-data concerning the antennas, sources, time observation periods and polarizations is captured for later processing in CASA [63].

The measurement set used for this work, is an observation of the Messier 83 (M83) galaxy corrupted by RFI. It is used in this research to show the effectiveness of the interference suppression algorithm. The data set was privately communicated by the KAT-7 staff who previously analyzed the data set manually.

For this reason, this section explains the composition of the measurement set, as well as shows the visualizations of the interference corrupted measurement set in various planes.

6.4.2 Measurement Set Format

The measurement set is a folder directory structure that nests tables
and meta-data concerning an observation in a single location. Each
measurement set contains information about various parameters of a
particular observation. These parameters as well as their descriptions may
be seen in Table IV.

TABLE IV

Fields in a Measurement Set [62]

Parameter	Description
ANTENNA1	First Antenna in Baseline
ANTENNA2	Second Antenna in Baseline
FIELD_ID	Celestial Source Identifier
DATA_DESC_ID	Description of the Data
ARRAY_ID	Sub-array Number
OBSERVATION_ID	Observation Identifier
POLARIZATION_ID	Polarization Identifier
SCAN_NUMBER	Scan Number
TIME	Observation Time of a Particular Source
UVW	(u, v, w) Coordinates of the Antennas

The radio astronomy suite CASA, allows for effective integration of the MS
directories in order to apply analysis and processing techniques. Tools such
as flagging and calibration are included in CASA and are easily applied to
measurement sets by adding columns to the existing tables.

As the wavelet based interference suppression system was implemented in
MATLAB, it was necessary to extract the time/frequency data from each
baseline at each polarization. This was done through the use of a *python*
script using that extracted the time-frequency observations for each baseline.
The number of baselines present in an array can be described as

$$N_{\text{baselines}} = \frac{N_{\text{ant}}(N_{\text{ant}} - 1)}{2}. \tag{6.13}$$

So that when using the 7 antennas in the KAT-7 array results in 21
different baselines at 4 different polarizations each with real and imaginary
components. This means that 168 .csv files were recovered for the wavelet
processing in MATLAB.

6.4.3 Messier 83 Galaxy

An observation of the Messier 83 (M83) galaxy was used for an application
of the implementation of the wavelet interference suppression system. The
recording of M83 was taken on 29th of July 2013 over a ten hour observation
period starting at 10:11am. An image obtained from an optical telescope
of the M83 galaxy can be seen in Figure 6.4a. It can be seen that M83 is a
barred-spiral galaxy, and it is known to be one of the closest and brightest
of its type to earth [4].

An image of M83 in the radio spectrum can be seen in Figure 6.4b,
it is the result of applying the CLEAN algorithm to the (l, m) plane
from the resultant 2D Fourier Transformed (u, v) plane for the Stokes
I polarization. The other polarizations of the image can be seen in
Appendix D.1.

It can be noted that the radio and optical spectra portray the same
galaxy significantly differently. In addition to this, it can be seen that there
is interference present in this radio image, as shown by the undulations in
the borders surrounding the source.

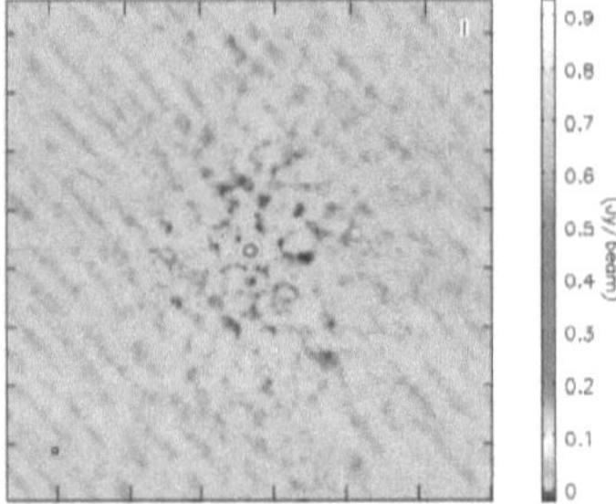

(a) Messier 83 Galaxy in the Visible
Spectrum [64]

(b) Messier 83 Galaxy in the Radio
Spectrum

Figure 6.4: Figures showing a Comparison of Messier 83 Galaxy in the
Visible and Radio Spectra

As the wavelet interference suppression algorithm has been shown to be
effective when applied to time domain data, it is necessary to apply it to the
time/frequency domain data obtained from each baseline. The associated
waterfall plot of the absolute value of the observation of M83 can be seen in

the left most plot of Figure 6.5. In this plot, various disturbances can be seen, most notably there is transient interference between channels 80 and 100. In addition to this, there is noticeable interference between channels 180 and 200 starting approximately at sample 700.

As the first channel's first stop-band is 1449.680 MHz and each channel's bandwidth is 781.250 kHz, it is possible to approximate the origins of the RFI present. For example, the RFI present in and around channel 88 is attributable to a 1.518 GHz satellite phone serviced by Inmarsat [65]. This procedure can be applied to validate that the disturbances present are in fact interference. However, as astronomical sources appear with far lower power levels, it can be assumed that all the high energy disturbances are interference [28].

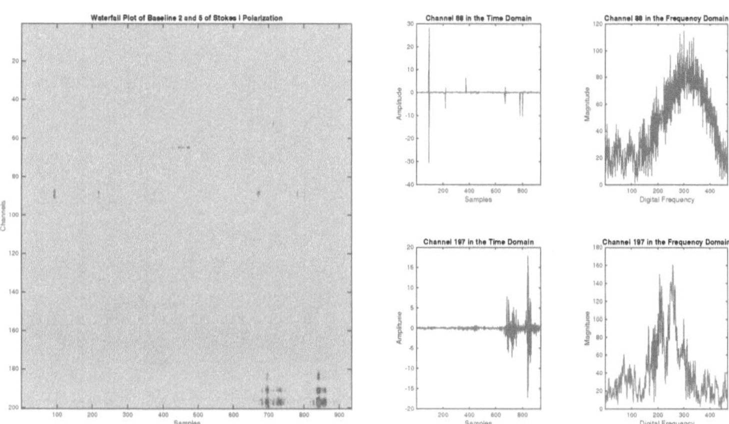

Figure 6.5: Figures showing the Time/Frequency Representation of the Baseline Pair Consisting of Antennas 2 and 5 for Stokes I

Although the waterfall plots of the baselines are insightful, more information can be obtained by examining the time and frequency domain characteristics of a particular channels of interest. The plot on the right hand-side of Figure 6.5 show the time and frequency domain waveforms for channels 88 and 197. The time domain representation of channel 88, shown in the top row of the figure, can be seen to be highly transient, containing sporadic impulses. Whereas channel 197's time domain representation, shown by the bottom row of plots of Figure 6.5, has transient interference located only at the end of the recording. Finally, it can be noted that both channel's frequency domain responses are broadband.

6.4.4 Data Flagging

As previously mentioned in Chapter 3, flagging is a common way of dealing with RFI in radio astronomy. In the case of this research project, it is useful to inspect the results of the flagging procedure such that comparisons can be drawn between the performance of the wavelet interference suppression algorithm and the flagged data. In this case, the flagging was performed manually by astronomers at the South African Radio Astronomy Observatory (SARAO) in order to be certain of the comparisons drawn.

In this dataset, flagging was performed in the time frequency plane, which resulted in masked regions that the clean algorithm would not use them to create the deconvolved image. The masked time/frequency plane can be seen in Figure 6.6.

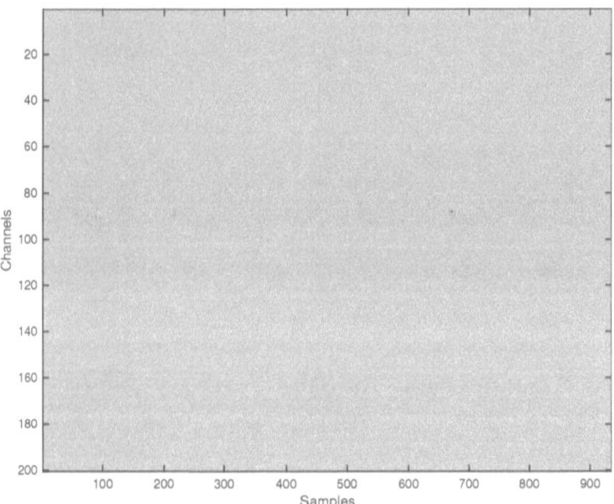

Figure 6.6: Masked Waterfall Plot of Baseline 2 and 5 for Stokes I Polarization

This image was obtained by taking the locations of the flags, and setting all corresponding samples in the time/frequency plot, to which a flag corresponds to, to zero. It can be seen that the transient interference that is present in Figure 6.5 has been attenuated significantly.

6.5 Interference Suppression

In the application of the interference suppression algorithm to radio astronomy the Messier 83 data set is used. The time domain vectors corresponding to each channel in the time/frequency plot are processed individually using the system implementation described in Section 5.5.

The preprocessing stage of the system implementation reads in each of the extracted .csv files from the MS and applies the detection and suppression stages as previously described. As the time/frequency representations of the recordings are in the form of complex integers, the wavelet based interference suppression algorithm must be applied to both the real and imaginary parts of each time vector separately.

In order to use the designed interference suppression system, models of the interference must be found as well as each interference model's corresponding wavelet. For this reason, this section documents this procedure.

6.5.1 Interference Modeling Algorithm

The interference modeling and extraction process used for the creation of the median based models of Messier 83 can be seen in flow chart in Figure 6.7. In order to create accurate models of the interference, the algorithm is applied to the real and imaginary components of each baseline at every polarization.

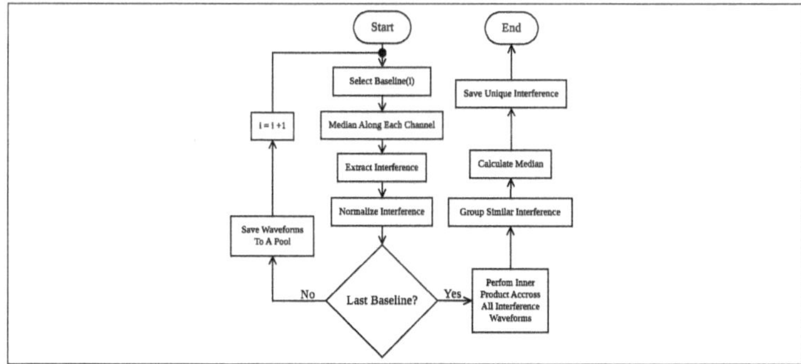

Figure 6.7: Flow Chart showing the Interference Extraction Algorithm

In the first stage of the algorithm, the median is taken along the frequency dimension of each baseline's time/frequency representation. This was done in order to reduce the amount of data to be processed by the automated RFI extraction algorithm. The median operation was performed piece-wise in time, such that the result was a single time-domain vector. The median operation allowed for easy detection of outliers, and the result of this applied to the baseline pair consisting of antennas 2 and 5 can be seen in Figure 6.8.

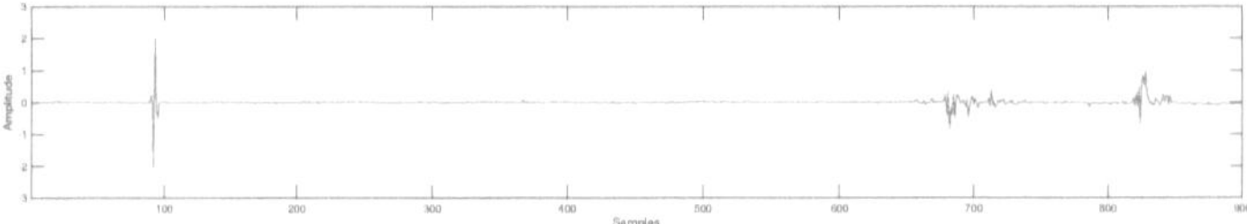

Figure 6.8: Median of Baseline Consisting of Antennas 2 and 5 for Stokes I Polarization

At this point, the automated RFI extraction algorithm is applied to index out each of the parts of the vectors corresponding to the interference. This process is applied until every baseline and every polarization is iterated over. The resultant interference waveforms are normalized and stored for later processing. Once all the interference is extracted, each of the saved waveforms is compared using the inner product.

The comparison is performed sequentially, such that on the first iteration, the first extracted waveform is compared to all others. If a waveform in the saved pool is similar to the selected interference waveform, then the waveform in the pool is removed and put into a matrix of similar interference waveforms. The algorithm then removes the next waveform from the pool and performs the same process until there is no interference left in the pool. This results in several matrices of grouped interference. The median of each of the matrices are then found and the resultant median-modeled waveforms are saved to file.

Once the models of the interference have been found, they are used to determine the optimal wavelets for the suppression system. These are determined using the entropy based methods discussed in Section 5.4. An illustration of the optimal wavelets corresponding to the modeled interference waveforms may be seen in Figure 6.9. In addition to this the median-based models are also used as the transfer function of the bank of matched filters used to detect each interference source.

It can be noted that the interference selection algorithm resulted in 13 different interference types, which in turn enabled the selection 13 matched

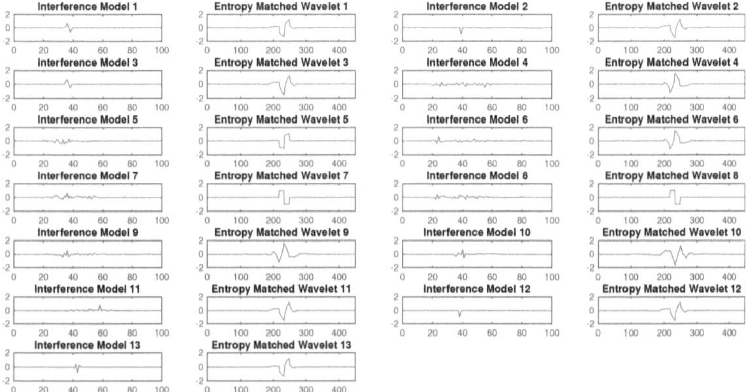

Figure 6.9: Plots showing the Modeled Interference and the Entropy Based Wavelets for M83

entropy wavelets. Furthermore, it can be noted that when the extracted modeled interference is of a complex shape, such as that shown by interference model 8, a suitable wavelet cannot be found and a Haar-like shape is assumed to be optimal. This is a limitation of the matching procedure used in this implementation, and it is a result of not using a higher dimensional library of wavelets.

6.5.2 Interference Suppression Algorithm

In Section 5.5, a formulation of the system was described that involved determining the precise location of interference in a vector consisting of multiple interference sources. This procedure was shown to be effective, however it is computationally costly when applying it to large data sets such as the astronomy measurement sets.

In addition to this, it was noted that the interference in each of the time domain vectors extracted from the measurement set only contained a single strong interference source. For this reason, the wavelet interference suppression algorithm was applied row by row to the time/frequency plots corresponding to each baseline and polarization. An overview of the algorithm applied to the astronomy data can be seen in Figure 6.10.

The algorithm receives the extracted `.csv` files corresponding to the real and imaginary components of each baseline at each polarization and sequentially processes each channel. The selected channel is then run through the bank of matched filters that are determined from the median based models. The matched filter with the maximum output indicates which source of interference is present in that channel and the matched filter's associated wavelet is used for the DWT decomposition.

Once the optimal wavelet has been chosen for the selected channel, it is then padded so that the resulting vector's length is a power of two. This is done because the filter-bank implementation of the DWT requires the input vector's length to be a power of two. Once the interference suppression has been performed, the resulting recovered channel is stripped of its padding samples and saved to file.

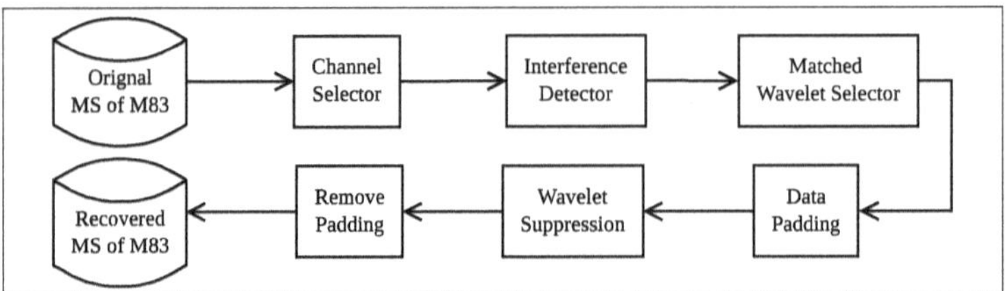

Figure 6.10: Block Diagram Illustration the Suppression Algorithm

6.5.3   Optimal Parameter Selection

The optimal parameters of wavelet based interference suppression algorithm are determined by considering the resultant power of the main source as well as the number of sources recovered from the algorithm. This these measurements are obtained through the use of Gaussian source modeling in a software package called Python Blob Detection and Source Finder (PyBDSF) [66].

The source counting is performed by modeling all peaks above a threshold in the deconvolved image as Gaussians. The threshold is determined by the regional RMS power of areas in the cleaned image. Then each of the detected peaks are grouped together by a proximity measure determined by the clean beam of the image.

Next the algorithm performs a grouping of the Gaussians into sources. The groups of islands are considered part of the same source if

- No pixels on the line joining the centers of any pair of Gaussians has a value less then the threshold.

- The line joining the centers of the Gaussians are separated by a distance less than the half of the sum of the FWHM.

In order to determine the received power of the source the fluxes of each of the grouped Gaussians are summed to obtain the total flux [66]. Using these methods, the power of the main source and the number of sources in an image is calculated for each decomposition level and for hard and soft thresholding. From this the optimal parameters to be used on the wavelet suppression system applied to M83 is calculated.

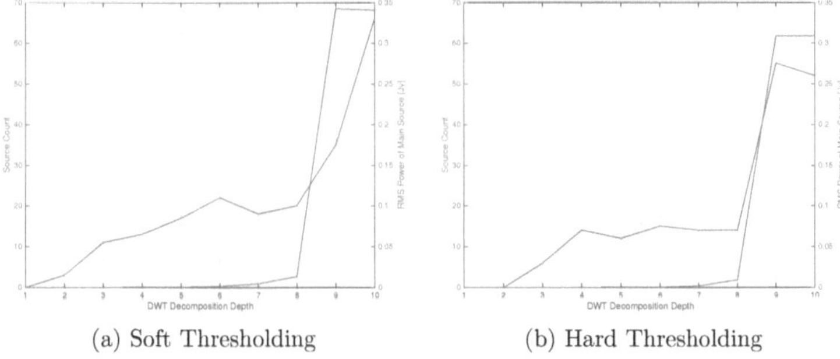

(a) Soft Thresholding　　　　(b) Hard Thresholding

Figure 6.11: Figures Showing the Relationship Between the Number of Decomposition Levels of the DWT with The Number of Extracted Sources and The RMS Power Level Of The Main Source for Hard and Soft Thresholding

Figure 6.11a shows the source count and the RMS power for each decomposition level using the universal adaptive threshold applied with soft thresholding. Whereas Figure 6.11b is the result of hard thresholding.

It can be seen that for both hard and soft thresholding the maximum source count is achieved when the decomposition depth, n, is 10, however soft thresholding yields a higher number of sources. For this reason soft thresholding is considered the optimal thresholding technique for M83.

Additionally, it can be seen that the decomposition level of 10 results in slightly lower power of the main source. As the source count for $n = 9$ is 35 and when $n = 10$ the source count is 66, the decomposition depth was set to 10 for the M83 data set.

Although it seems as though the source count and power of the main source increases as n get greater, it is not possible to increase $n > 10$. This is because the maximum number of decomposition levels is equal to $log_2(L)$, where L is the length of the input sequence.

From the design of the interference suppression algorithm and the selection of optimal parameters, results can be shown and conclusions can be made about the effectiveness of the system.

Chapter 7

Results

This chapter considers the design and application of the wavelet based interference suppression system to radio astronomy data and shows the respective results of the system. This is achieved by showing the sensitivity and accuracy of the system in both the context of radio astronomy as well as an application agnostic manner.

7.1 Background to Experimentation

The results obtained from the design and implementation of the wavelet based suppression system shown in Chapters 6 and 5 are documented in this section. In order to effectively describe the suppression scheme's performance this chapter segments the results into two different sections. These being, the application non-specific performance of the suppression scheme and the results obtained from the astronomical application of the system.

The application agnostic results of the wavelet based interference suppression scheme are those that do not require astronomy based techniques to measure performance. These results can be described by the suitability of wavelets in the library, the worst case scenario power loss and the computational complexity associated with the suppression system. Corollary, the application specific results are obtained using astronomy based techniques.

In order to ensure statistical significance of each of the results, each experiment was run 100 times unless otherwise specified.

7.2 Wavelet Based Interference Suppression System Results

7.2.1 Matched Wavelet Selection Similarity

In order to document the matched wavelets' similarity to their modeled interference waveforms, measures of entropy and RMSE are used as documented by Chapter 5. This is achieved by iteratively selecting each wavelet and interference pair and computing the DWT when -5dB of noise is added. The entropy of the decomposition is then measured at the analysis stage of the DWT and the resultant interference free synthesized waveform's RMSE is computed.

The system parameters determined in the previous section were used so that the number of decomposition levels is 10 and the threshold was performed using soft thresholding methods. This result may be seen in Figure 7.1.

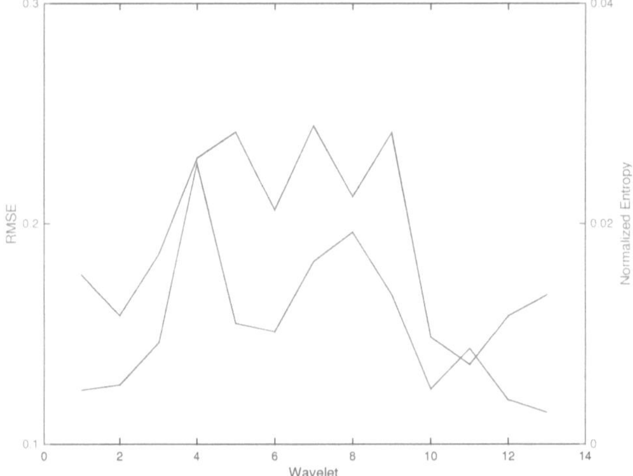

Figure 7.1: Figure Showing the Resultant Entropy and RMSE for Each Wavelet In the Library

It can be seen that there is a correlation between the entropy of each

104

wavelet and its associated RMSE. However for wavelet numbers four to nine, there seems to be a slight offset between the entropy and RMSE lines. Furthermore it can be noted that entropy of each wavelet is below 0.04, which in comparison to Figure 5.9 suggests that the matches are strong. Finally it can be seen, that the resultant RMSE is significantly higher than the Figure 5.9, which may be attributable to the -5dB noise level.

7.2.2 Sensitivity Analysis

The sensitivity of the wavelet based suppression system is tested by considering the worst case power loss scenario. In all cases, the worst case power loss is attributable to the over suppression of the input signal. This is as a result of the threshold being set too low, such that the system excises parts of the signal of interest that should be preserved.

Therefore, when a vector containing only Gaussian noise is injected into the system, the observed power loss is indicative of the worst case functioning of the system. In this test, -5dB of noise was injected into the system and the DWT depth was fixed to 10 decomposition levels using soft thresholding. Figure 7.2a shows the associated RMS power loss percentage for each of the wavelets used in the astronomy based application.

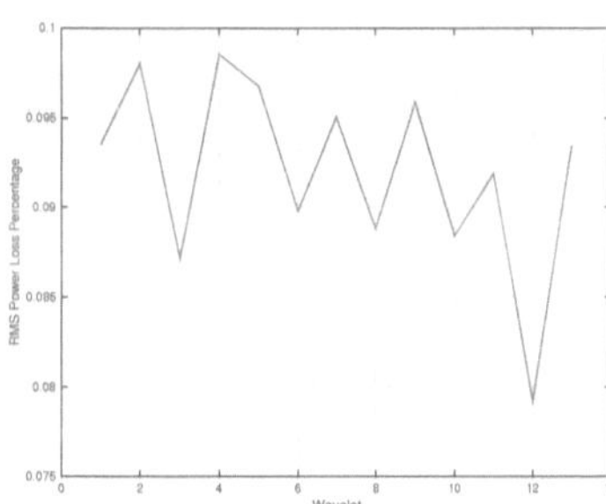

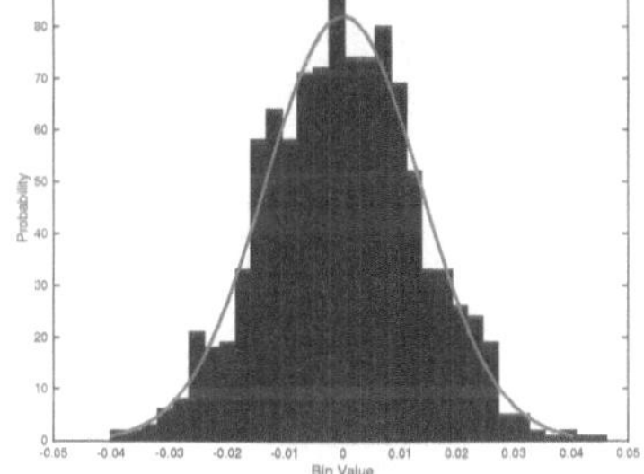

(a) Figure Showing the Worst Case Average RMS Power Loss Associated with Each Wavelet in the Designed Library

(b) Figure Showing the Histogram Resulting From the Average Worst Case Power Loss of Each Wavelet in the Designed Library

Figure 7.2: Figures Showing the Sensitivity of the Wavelet Suppression Algorithm when -5dB of Gaussian White Noise is Injected into the Input

It can be seen that the forth wavelet results in a 10% power loss and is the upper-limit of measured percentage RMS power loss. Additionally, the twelfth wavelet has the best preservation of power when the input is Gaussian. It can be said from these results that the expected worst case RMS power loss is between 8% and 10%.

Furthermore, Figure 7.2b proves that in the worst case scenario, the output of the system preserves the Gaussian distribution of the input. The concatenated output of the 100 iterations of the sensitivity tests all do not reject the null hypothesis of the Lilliefors test.

7.2.3 Time Complexity

The time complexity results are obtained by testing the two subsystems of the wavelet based suppression algorithm. These being the interference detection and interference suppression subsystems. As the M83 data set is 935 samples long in the temporal dimension, a 1024 length vector of ones is used for testing the time complexity of the system. The time complexity tests performed in this section were executed on the server specified in Listing 4.3.4.

Furthermore, in order to show the general temporal performance of the system, each of the two subsystems were evaluated at decomposition levels ranging from 1 to 10. This was done to document the performance of the system if it were to be applied to a data set other than the M83 data set that used a decomposition level of 10. The results obtained from these tests can be seen in Figure 7.3.

It can be seen that as the number of decomposition levels increase so does the time complexity of the suppression subsystem. In addition to this it can be seen that the matched filtering or interference detection stage has a constant time complexity. This result is expected as the number of decomposition levels of the DWT should have no effect on the matched filtering operations performed in the interference detection stage of the algorithm.

In addition to this, as the M83 dataset consists of $168\ 200{\times}935$ matrices and the suppression is performed down the frequency dimension of the matrices. This means that the suppression system has to compute approximately 30 million suppressions. For this reason the overall time taken of the system

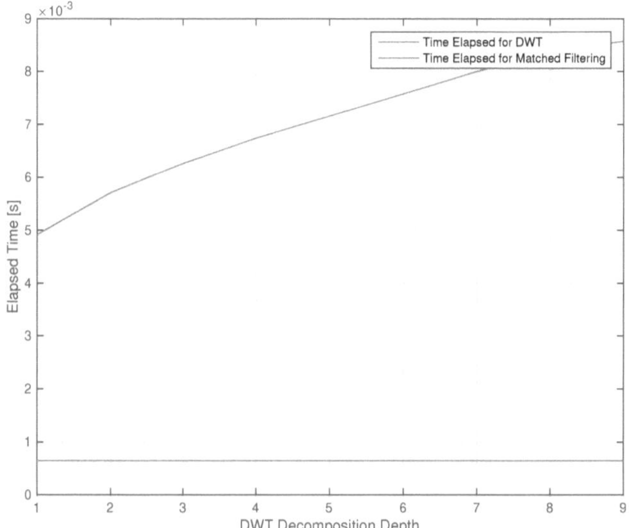

Figure 7.3: Figure Showing the Relationship Between the Number of Decomposition Levels of the DWT and the Time Complexity of the Wavelet Suppression System. It can be Seen That as the Number of Decomposition Level Increase so Does the Time Taken to Perform the Suppression.

when applied to the M83 dataset is approximately 5 minutes.

7.3 Radio Astronomy Specific Results

Using the implementation described in Section 6.5, the following radio astronomy specific results are made. In order to effectively document the obtained results, this section is segmented into two different sections. Namely the results obtained through the use of imaging and the CLEAN algorithm as well as the statistical robustness of the resultant suppression using the wavelet based interference suppression scheme.

7.3.1 Imaging Results

The Stokes I polarization of the deconvolved images obtained from the CLEAN algorithm for the original, flagged and recovered measurement sets can be seen in Figure 7.4. The parameters used in the CLEAN algorithm are shown in Table V. These parameters were kept consistent across all images shown in order to ensure comparability of the results.

In the original interference corrupted image, shown by Figure 7.4a, it can be seen that there are periodic undulations running diagonally around the main source in the center of the image. In addition to this it can be seen that the some sources seem to overlap with the interference. The other images corresponding to all other polarizations of the original interference corrupted M83 dataset can be seen in Appendix D.1.

TABLE V

TABLE SHOWING THE CLEAN USED TO IMAGE M83

CLEAN Parameter	Value
Number of Iteration	10000
Image Size	1024x1024 pixels
Cell Size	0.25 arcmin
Weighting	Briggs
Robust	-2

Figure 7.4b shows the resultant image associated with the manually flagged data. It can be seen that the interference is no longer present in the image and the sources that are outlined by the white contours are significantly more visible. In addition to this the other polarizations of the flagged images can be seen in Appendix E.1.

Finally, Figure 7.4c shows the image that corresponds to data which the interference has been suppressed from using the wavelet based methods. It can be seen there are still slight undulations present, however these are attributable to the cleaning algorithm not cleaning deep enough. However for purposes of comparison with the original data the CLEAN parameters were fixed.

In addition to visual inspection of the imaged data, the source count and and main source power are calculated, as shown in Table VI. It can be seen that the source count increases from 48 to 59 when flagging is applied to

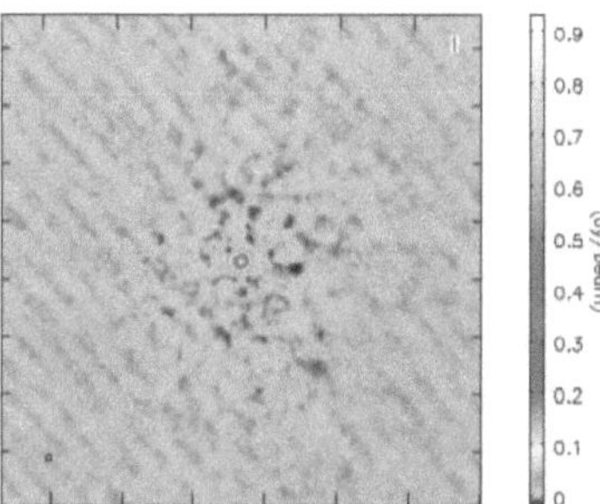

(a) Image of M83 from RFI Corrupted
Data

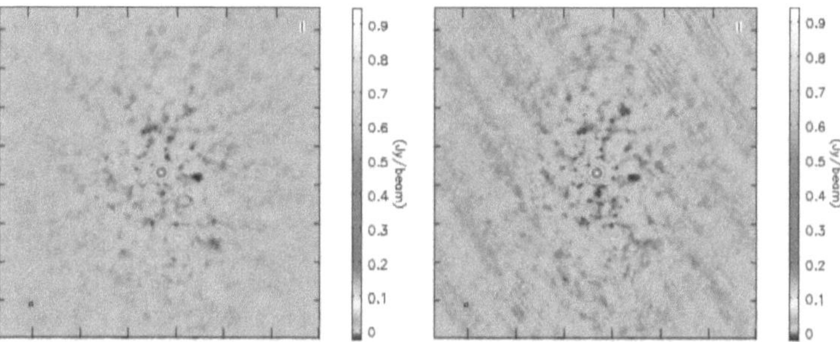

(b) Image of M83 from Flagged Data   (c) Image of M83 from Recovered Data

Figure 7.4: Images showing the Results of Flagging and Wavelet Based
Suppression on Stokes I Polarization

the original data and when the wavelet based interference mitigation scheme
is applied, the source count increases to 66. The wavelet based interference
suppression algorithm results in a 27% increase in the number of sources from
the original, and has a 10% increase from the flagging algorithm.

Finally, the power of the main source increases from flagging by 1.5% and
there is a 1.9% loss in power of the main source between the flagging data
and the recovered data.

TABLE VI

Table Showing a Comparison of the Source Count and Main Source Power for Each Image

Image Type	Source Count	Main Source Power [Jy]
Original Image	48	3.821538×10^{-1}
Flagged Image	59	3.881248×10^{-1}
Recovered Image	66	3.807224×10^{-1}

7.3.2 Statical Robustness

As astronomical, interference free, data is normally distributed, it is useful to analyze the output distributions before and after interference suppression. From this reasoning it is possible to validate whether the interference suppression algorithm is effective. Figure 7.5 shows each of the histograms of the original, flagged and interference suppressed data sets.

The histograms are formed by flattening each of the complex samples and finding their magnitude. This is followed by a normalization to ensure that each of the means of each of the baselines are centered at zero. Each of the flattened normalized magnitude vectors are then concatenated together and the histogram is plotted. The solid blue bars correspond to each of the bins of the histogram and the red line is the best fit histogram to the distribution as performed by the MATLAB `histfit` function.

It can be seen that the distribution of the original interference corrupted data set is clearly not Gaussian, as shown by Figure 7.5a. It can be seen that there are many outliers extending from the bin value of -200 to 100. This result is validated by the original distribution failing the Lilliefors test.

The resultant flagged histogram, shown by Figure 7.5b, shows that flagging improves the distribution to appear more Gaussian. However there are still outliers and there is a large peak around the zeroth bin. The flagged distribution also fails the Lilliefors test and for this reason it can be concluded that it is not Gaussian

Finally it can be seen that the wavelet based suppressed data set resembles the best normal distribution of the three, as shown by Figure 7.5c. It can be seen that there are fewer outliers and that the peak surrounding the zeroth bin is significantly lower than that of the flagged and original data sets. In addition to this, the recovered distribution fails the Lilliefors test.

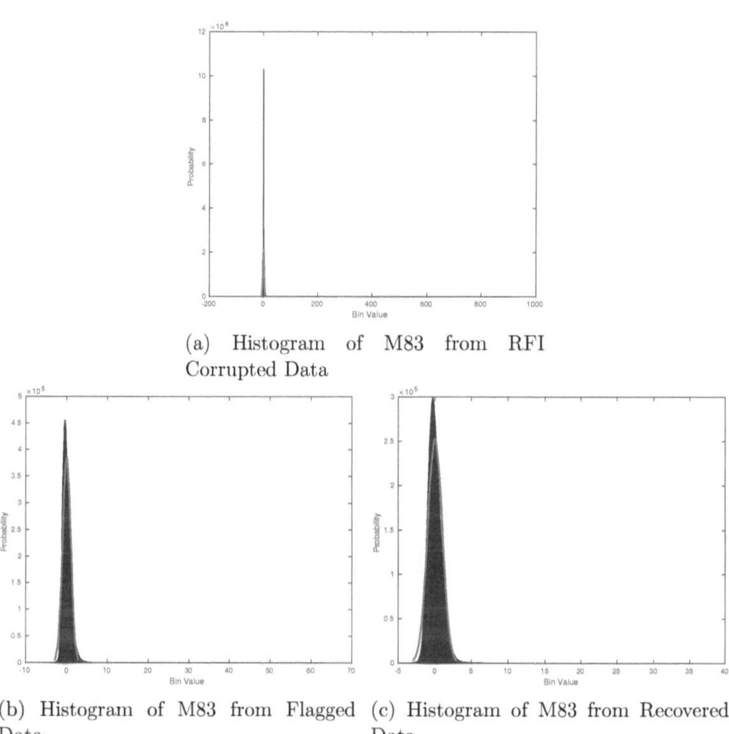

(a) Histogram of M83 from RFI Corrupted Data

(b) Histogram of M83 from Flagged Data

(c) Histogram of M83 from Recovered Data

Figure 7.5: Figures showing the Histograms of the Result of Flagging and Wavelet Based Suppression on the M83 Data Set

Chapter 8

Conclusions

In sight of the fact that this project aimed to design and implement a wavelet based interference suppression system, it is necessary to elaborate on the successes and failures of the designed system. The results obtained in Chapter 7 act as a heuristic of performance of the overall system and give insight into the the trade-offs involved with the development of the interference suppression system. Based on the results obtained in Chapter 7, the following conclusions may be drawn.

8.1 Improvements to Wavelet Matching and Interference Modeling

It can be concluded that improvements can be made to the multi-dimensional wavelet library. Through the visual inspection of Figure 6.9, it can be seen that there are wavelets in the designed library that uncorrelated to the modeled interference. This is confirmed by Figure 7.1 that there are insufficient similarities between the interference waveforms and their corresponding wavelets. This is attributable to the fact that the dimensionality of the library is too low.

In addition to this, it can be noted that although the smoothness condition ensures no discontinuous artefacts in the resynthesized waveforms it may not be optimal for matching interference to wavelets. Considering the RFI present in Figure. 6.5, it can be seen that the interference waveforms are jagged discontinuous. The effect of this is that the performance of the wavelet based interference suppression system may be improved by allowing a subclass of discontinuous wavelets into the interference suppression library.

Furthermore, the modeling of the interference waveforms is suboptimal. In research by Czech et al. [26] it is shown more robust estimates of interference can be obtained through the use of advanced statistical methods. This in turn would allow for more accurate wavelet matching to be performed, thereby increases the sensitivity and performance of the wavelet interference suppression system.

8.2 Inadequate Power Loss Statistics

The worst case RMS power loss of 8% to 10% attributed to the suppression system is unavoidable. This is because the implementation of adaptive wavelet coefficient thresholding is agnostic to the input data stream. As previously explained, the optimal results are obtained when there is a perfect match between the wavelet used for the decomposition and the input signal. In the astronomy based application it is shown that interference is effectively suppressed which is directly as a result of the use of suitable wavelets. This being said, the 1.9% power loss is a result of the over suppression of a number of the time domain vectors from each baseline. This may be rectified by using higher dimensional wavelets in the matched library or by improving the models used in the detection stage of the implementation.

For this reason, it can be concluded the associated power loss is unavoidable, however with better designed wavelets and interference models a lower power loss percentage may be obtained.

8.3 Suitable Time Complexity

Computational complexity is a important result of the wavelet based interference suppression system as it dictates suitability to many different use cases. This being said, as the objective of this research project was to design and validate a wavelet based interference suppression scheme that can be used for astronomy purposes, little effort was made to optimize the temporal efficiency of the system.

It must be noted that the application of the wavelet suppression scheme is an embarrassingly parallel problem. As each of time vectors are completely independent to one and other. The effect of this is that the interference in each of the baselines at every polarization may be suppressed independently to each other. This parallel implementation may be performed on a variety of architectures such as the FPGA's on site at the SKA facilities.

For these reasons, it can be concluded that the time complexity is adequate, however there is undoubtedly means of improving the obtained results.

8.4 Performance Comparison of Flagging and the Suppression System

The obtained results concerning the number of sources in the radio astronomy images of the Messier 83 Galaxy prove that the wavelet based suppression system is superior to flagging techniques. Albeit at the cost of a slight decrease in power of the main source. The fact that the wavelet based suppression system is automated is a great advantage of the system. For example, if one considers a similar observation duration as the one performed in the Messier 83 MS, but uses the MeerKAT project with 64 antennas [67] then there are 2016 different baselines. This in turn would result in a 3000% increase in data, which would become exceptionally challenging to flag by hand.

For this reason it can be concluded that the wavelet based suppression system is highly effective in aiding astronomers in the mitigation of interference.

8.5 Statistical Analysis

The results shown by Figure 7.5 clearly show the effectiveness of the wavelet suppression scheme and give insight into reason why there is an increase in number of sources from flagged data. The failure to pass the Lilliefors may be attributable to the fact that this particular observation of the M83 Galaxy may have intrinsic non-linear effects over the observation. This being said, it can be concluded that the wavelet based suppression scheme is statistically

superior due to the fact that there is a clear reduction of outliers between
the flagged histogram and the histogram as a result of wavelet suppression.

Chapter 9

Recommendations and Future Work

As a consequence of the conclusions drawn, it is natural to describe how the designed wavelet based suppression system could be improved. Therefore this final chapter will discuss the further advancement of the system in order to make it capable of being used in real world scenarios.

9.1 Improvements to Interference Detection

As there was a stronger focus of this research project on wavelet interference suppression, there room to improve on the interference detection front. Although it has been proven that the techniques used have been effective, alternative techniques should be considered. For this reason, three different approaches to the classification and detection of interference have been considered for the future work of this project.

9.1.1 Classification in the Wavelet Domain

The wavelet transform gives insight into the scaling phenomena of sources, therefore it could be used as a means of classification of different sources. This could be achieved using methods similar to those used by Czech et al. [26], through the modeling of characteristic features in the wavelet domain of a number of transients. Such that when a decomposition is done to a particular time domain vector, the match to a particular scale could be used as an indicator of the presence of the signal.

9.1.2 Physical Characterization of Sources

Methods of classification could be used for automatically classifying sources similar to that done in a paper by Doran [25]. Where physical attributes such as time of day and the orientation of the telescopes were used in the classification of interference in radio astronomy.

9.1.3 Machine Learning Techniques

From the shown in the Chapter 3, there have been significant improvements of the use of machine learning techniques for the classification of interference for radio astronomy [25, 26, 32]. For this reason, an interesting recommendation for the future work of this project could be the integration of classification tools such as Artificial Neural Networks (ANN) and clustering algorithms that are used in the machine learning field.

Through the use of these sophisticated tools, the system could be applied more freely to other sorts of data sets and the modeling and wavelet matching could be done autonomously by the machine learning system.

9.2 Wavelet Based Considerations

The wavelet based suppression scheme has been shown to be effective. However as mentioned in Chapter 8 improvements can be made on the choice of wavelet as well as the scheme in which the DWT is implemented.

9.2.1 Higher Dimensional Wavelet Libraries

A key feature of the multi-dimensional library used for the excision of interference is that the wavelets it includes do not have a high number of zero crossings. Therefore for more complex interference shapes, it becomes challenging to find an optimal wavelet in lower dimensional libraries. For this reason, higher dimensional wavelet library creation should be used to obtain more complex wavelet shapes. A trade off of this would be the fact that the higher the dimension of wavelets, the more taps each of the wavelet filters would have. Which in turn would increase the computational complexity of the DWT thereby causing an increased computation time.

9.2.2 Alternative Wavelet Architectures

In order to give a more holistic overview of the wavelet interference suppression landscape, it is recommended that this project investigate alternative wavelet architectures. For example the benefit of using the wavelet packet decomposition is that better basis may be found for the representation of interference. Additionally, the wavelet lifting scheme has been shown to yield exceptionally fast computations times thereby making it a useful implementation of a real time interference suppression algorithm. Alternatively, biorthogonal wavelets could be used in pursuit of finding stronger matches to interference.

In addition to this, a more sophisticated wavelet design scheme that incorporates wavelets that are more discontinuous into the library. This could be achieved through the relaxation of the smoothness criterion strongly enforced in the design of the smooth, unique wavelet library.

9.3 Implementation Based Considerations

The wavelet based interference suppression scheme has several limitation. In order to address these limitations suggestions are made so that this project can be used to solve problems in future applications.

9.3.1 System Limitations

A major limitation of the wavelet based suppression scheme is that the DWT requires the input vector's size to be a power of two. Classically this is addressed in other implementations by zero padding or by re-sampling the vector [2]. In the system's current implementation mean based padding is performed such that samples with a magnitude equivalent to the mean of the input vector are appended. This has been shown to be effective when is input vector's length is near to a power of two.

However, considering the case that the input vector length is 4500, the next power of two is 8192 which means that the input vector almost has to be doubled in size so that the DWT can be performed. The effect of this is that the adaptive threshold will be set too low and data will be lost.

A possible solution for larger data sets is windowing out sections of the signal that are powers of two and performing the DWT in batches as done in Hazas's Master book [20].

9.3.2 Integration into Existing Technology

A further limitation of the wavelet suppression system is its lack of integration into existing radio astronomy based infrastructure. As the system stands, the measurement sets need to be extracted into `.csv` file formats in order to perform the suppression. In the field of radio astronomy, commonly used techniques such as the CLEAN algorithm and data visualizations have been integrated into the CASA software suite [62]. For this reason it seems appropriate for the wavelet based suppression scheme to also be integrated into CASA.

Additionally, an effective way to integrate the wavelet based suppression scheme would be to incorporate it into the FPGA based front end on the SKA site. This would also allow for native parallelization as FPGA's are well suited to create parallel implementations of algorithms.

9.4 Application to a Variety of Data Sets

A reasonable recommendation that can be made for the designed wavelet suppression scheme is attempting to apply it to different data sets. Furthermore, as a multitude of scientific fields suffer for the effects of interference it would be useful in future work to investigate other applications.

9.4.1 Alternative Astronomy Data Sets

In terms of astronomical data sets, it would be insightful to apply the wavelet based interference suppression system to other measurement sets. Namely those that contain different celestial sources that have significantly different signatures. For example pulsars emit electromagnetic radiation bursts periodically, which may be detected as interference by the wavelet based suppression system.

9.4.2 Electrical Engineering Applications

There are several application of the interference suppression algorithm in the field of electrical engineering, the most notable are telecommunications as well as radar technologies. In the case of power line communications, packets of information are transferred over power lines in order to decrease the amount of infrastructure needed to deploy a network to under-served communities. A consequence of a network like this is that the long power lines are highly susceptible to transient electromagnetic interference from high voltage sources such as Compact Fluorescent Lights (CFL) and circuit breakers. Therefore it would be useful to apply these wavelet based techniques to power line communications.

Another application of the wavelet based suppression system is to that of passive radars, as they experience a phenomenon called feed through. Passive radar is performed by using one antenna, called the reference antenna, to monitor continuously transmitting source such as a Frequency Modulated (FM) radio station and using a secondary antenna, referred as the surveillance antenna, to monitor an area of the sky. This configuration of a passive radar receives reflected electromagnetic radiation reflections from targets such as planes and is able to monitor its position and velocity.

The feed through issue arises when a portion of the reference signal interferes with the signal received by the surveillance antenna. For this reason, passive radar feed through suppression is a well suited problem to wavelet based interference suppression.

9.5 Closing Comments

Although many recommendations have been made to improve the performance and application of the wavelet based suppression scheme. It must be said that the designed system has accomplished the objectives listed in Chapter 1 and for this reason is successful.

References

[1] A. R. Foley, T. Alberts, R. P. Armstrong, A. Barta, E. F. Bauermeister,
H. Bester, S. Blose, R. S. Booth, D. H. Botha, S. J. Buchner,
C. Carignan, T. Cheetham, K. Cloete, G. Coreejes, R. C. Crida,
S. D. Cross, F. Curtolo, A. Dikgale, M. S. de Villiers, L. J. du Toit,
S. W. Esterhuyse, B. Fanaroff, R. P. Fender, M. Fijalkowski, D. Fourie,
B. Frank, D. George, P. Gibbs, S. Goedhart, J. Grobbelaar, S. C.
Gumede, P. Herselman, K. M. Hess, N. Hoek, J. Horrell, J. L. Jonas,
J. D. Jordaan, R. Julie, F. Kapp, P. Kotzé, T. Kusel, A. Langman,
R. Lehmensiek, D. Liebenberg, I. J. Liebenberg, A. Loots, R. T. Lord,
D. M. Lucero, J. Ludick, P. Macfarlane, M. Madlavana, L. Magnus,
C. Magozore, J. A. Malan, J. R. Manley, L. Marais, N. Marais, S. J.
Marais, M. Maree, A. Martens, O. Mokone, V. Moss, S. Mthembu,
W. New, G. D. Nicholson, P. C. van Niekerk, N. Oozeer, S. S. Passmoor,
A. Peens-Hough, A. B. Pińska, P. Prozesky, S. Rajan, S. Ratcliffe,
R. Renil, L. L. Richter, D. Rosekrans, A. Rust, A. C. Schröder,
L. C. Schwardt, S. Seranyane, M. Serylak, D. S. Shepherd, R. Siebrits,
L. Sofeya, R. Spann, R. Springbok, P. S. Swart, V. L. Thondikulam,
I. P. Theron, A. Tiplady, O. Toruvanda, S. Tshongweni, L. van den
Heever, C. van der Merwe, R. van Rooyen, S. Wakhaba, A. L. Walker,
M. Welz, L. Williams, M. Wolleben, P. A. Woudt, N. J. Young, and
J. T. Zwart, "Engineering and science highlights of the KAT-7 radio
telescope," *Monthly Notices of the Royal Astronomical Society*, vol. 460,
no. 2, pp. 1664–1679, 2016.

[2] G. Strang and N. Truong, *Wavelets and Filter Banks*, 2nd ed., 1997.
[Online]. Available: http://www.amazon.ca/exec/obidos/redirect?tag=
citeulike09-20{%}5C{&}path=ASIN/0961408871

[3] I. Daubechies and W. Sweldens, "Factoring wavelet transforms into
lifting steps," *The Journal of Fourier Analysis and Applications*,

vol. 4, no. 3, pp. 247–269, 1998. [Online]. Available: http://link.springer.com/10.1007/BF02476026

[4] H. Frommert and C. Kronberg, "Messier 83," 2008. [Online]. Available: http://www.messier.seds.org/m/m083.html

[5] I. Daubechies, *Ten Lectures on Wavelets.* Society for Industrial and Applied Mathematics, jan 1992. [Online]. Available: http://epubs.siam.org/doi/book/10.1137/1.9781611970104

[6] R. Merry and M. Steinbuch, "Wavelet theory and applications," *Literature Study, Eindhoven University*, p. 41, 2005. [Online]. Available: http://alexandria.tue.nl/repository/books/612762.pdf

[7] S. Mallat, "A Wavelet Tour of Signal Processing," *A Wavelet Tour of Signal Processing*, 2009.

[8] M. Misiti, Y. Misiti, G. Oppenheim, and J.-M. Poggi, "Wavelet Toolbox ™ User's Guide How to Contact MathWorks," p. 700, 2015. [Online]. Available: http://www.mathworks.com/help/pdf{_}doc/wavelet/wavelet{_}ug.pdf

[9] M. T. Hanna and S. A. Mansoori, "The discrete time wavelet transform: Its discrete time Fourier transform and filter bank implementation," *IEEE Transactions on Circuits and Systems II: Analog and Digital Signal Processing*, vol. 48, no. 2, pp. 180–183, 2001.

[10] K. I. Soman, K. P., Ramachandran, *Insight Into Wavelets : from Theory to Practice*, 3rd ed. Asoke K. Ghosh, 2010. [Online]. Available: https://books.google.co.za/books?id=6so1{%}5C{_}nCQl-AC

[11] P. Cavalier and D. O. Hagan, "A New Potential Field Shape Descriptor Using Continuous Wavelet Transforms," *Geophysics (in review)*, no. October, pp. 1–12, 2018.

[12] C. S. Burrus, *Tutorials on Wavelets and Wavelet Transforms*, C. S. Burrus, Ed. OpenStax-CNX, 2015.

[13] C. S. Burrus, R. a. Gopinath, and H. Guo, *Introduction to Wavelets and Wavelet Transforms: A Primer*, 1998. [Online]. Available: http://www.amazon.com/Introduction-Wavelets-Wavelet-Transforms-Primer/dp/0134896009

[14] R. A. Gopinath, J. E. Odegard, and C. S. Burrus, "Optimal Wavelet Representation of Signals and the Wavelet Sampling Theorem," *IEEE*

Transactions on Circuits and Systems II: Analog and Digital Signal Processing, vol. 41, no. 4, pp. 262–277, 1994.

[15] D. L. Donoho and I. M. Johnstone, "Ideal spatial adpatation by wavelet shrinkage," *Biometrika*, vol. 81, no. June 1992, pp. 425–455, 1994.

[16] D. L. Donoho, "De-Noising by Soft-Thresholding," *IEEE Transactions on Information Theory*, vol. 41, no. 3, pp. 613–627, 1995.

[17] D. Donoho and I. Johnstone, "Threshold selection for wavelet shrinkage of noisy data," *Proceedings of 16th Annual International Conference of the IEEE Engineering in Medicine and Biology Society*, pp. A24–A25, 1994. [Online]. Available: http://ieeexplore.ieee.org/lpdocs/epic03/wrapper.htm?arnumber=412133

[18] R. Cohen, "Signal Denoising Using Wavelets," no. February, pp. 1–29, 2012. [Online]. Available: http://tx.technion.ac.il

[19] W. I. M. Sweldens, "The Lifting Scheme: A Construction of Second Generation Wavelets," *SIAM J. MATH. ANAL.*, vol. 29, no. 2, pp. 511–546, 1998.

[20] M. Hazas and H. Hall, "Processing of Non-Stationary Audio Signals," Ph.D. book, 1999.

[21] R. R. Coifman and M. V. Wickerhauser, "Entropy-based algorithms for best basis selection," *IEEE Transactions on Information Theory*, vol. 38, no. 2, pp. 713–718, 1992.

[22] J. Lahtinen, J. Uusitalo, T. Ruokokoski, and J. Ruoskanen, "Evaluation and comparison of RFI detection algorithms," *14th Specialist Meeting on Microwave Radiometry and Remote Sensing of the Environment, MicroRad 2016 - Proceedings*, pp. 62–67, 2016.

[23] D. Czech, A. Mishra, and M. Inggs, "Characterizing transient radio-frequency interference," *Radio Science*, vol. 52, no. 7, pp. 841–851, 2017.

[24] H. W. Lilliefors, "On the Kolmogorov-Smirnov Test for Normality with Mean and Variance Unknown," *Journal of the American Statistical Association*, vol. 62, no. 318, p. 399, jun 1967. [Online]. Available: https://www.jstor.org/stable/2283970?origin=crossref

[25] G. Doran, "Characterizing Interference in Radio Astronomy Observations through Active and Unsupervised Learning," no. August, 2012.

[26] D. Czech, A. Mishra, and M. Inggs, "A Dictionary Approach to Identifying Transient RFI," *Radio Science*, vol. 53, no. 5, pp. 656–669, 2018.

[27] J. R. Fisher, "RFI and How to Deal with It," in *Single-Dish Radio Astronomy: Techniques and Applications*, vol. 278, 2002, pp. 433–445.

[28] J. M. Ford and K. D. Buch, "RFI mitigation techniques in radio astronomy," in *2014 IEEE Geoscience and Remote Sensing Symposium.* IEEE, jul 2014, pp. 231–234. [Online]. Available: http://ieeexplore.ieee.org/document/6946399/

[29] A. R. Millenaar and A.-j. Boonstra, "The Radio Frequency Interference Environment At Candidate SKA Sites," 2010.

[30] D. A. Mitchell and J. G. Robertson, "Reference antenna techniques for canceling radio frequency interference due to moving sources," *Radio Science*, vol. 40, no. 5, pp. 1–9, 2005.

[31] K. D. Buch, K. Naik, S. Nalawade, S. Bhatporia, Y. Gupta, and A. B., "Implementing and Characterizing Real-Time Broadband RFI Excision System for the GMRT Wideband Backend," *IETE Technical Review (Institution of Electronics and Telecommunication Engineers, India)*, vol. 4602, pp. 1–9, 2018. [Online]. Available: https://doi.org/10.1080/02564602.2018.1450650

[32] C. J. Wolfaardt, "Machine learning approach to radio frequency interference(RFI) classification in radio astronomy," no. March, p. 88, 2016.

[33] R. Lord and M. Inggs, "Efficient RFI suppression in SAR using a LMS adaptive filter with sidelobe suppression integrated with the range-Doppler algorithm," in *IEEE 1999 International Geoscience and Remote Sensing Symposium. IGARSS'99 (Cat. No.99CH36293)*, vol. 1, no. 5. IEEE, 1984, pp. 574–576. [Online]. Available: http://ieeexplore.ieee.org/document/773569/

[34] T. Witaszczyk and R. van Nieuwpoort, "Radio frequency interference mitigation for software telescopes." Ph.D. book, 2010.[Online]. Available: http://www.few.vu.nl/{~}rob/masters-theses/Tomasz-Wwitaszcsyk.pdf

[35] Y. Cendes, P. Prasad, A. Rowlinson, R. A. Wijers, J. D. Swinbank, C. J. Law, A. J. van der Horst, D. Carbone, J. W. Broderick, T. D. Staley, A. J. Stewart, F. Huizinga, G. Molenaar, A. Alexov, M. E.

Bell, T. Coenen, S. Corbel, J. Eislöffel, R. Fender, J. M. Grießmeier, P. Jonker, M. Kramer, M. Kuniyoshi, M. Pietka, B. Stappers, M. Wise, and P. Zarka, "RFI flagging implications for short-duration transients," *Astronomy and Computing*, vol. 23, pp. 103–114, 2018. [Online]. Available: https://doi.org/10.1016/j.ascom.2018.04.001

[36] S. Maslakovic, I. Linscott, M. Oslick, and J. Twicken, "A Library-Based Approach to Design of Smooth Orthonormal Wavelets," in *Proceedings of the 1998 IEEE International Conference on Acoustics, Speech and Signal Processing, ICASSP '98 (Cat. No.98CH36181)*, vol. 3. IEEE, 1998, pp. 1793–1796. [Online]. Available: http://ieeexplore.ieee.org/document/681808/

[37] V. S. Chourasia and A. K. Tiwari, "Design Methodology of a New Wavelet Basis Function for Fetal Phonocardiographic Signals," *The Scientific World Journal*, vol. 2013, pp. 1–12, 2013. [Online]. Available: http://www.hindawi.com/journals/tswj/2013/505840/

[38] J. O. Chapa and R. M. Rao, "Algorithms for designing wavelets to match a specified signal," *IEEE Transactions on Signal Processing*, vol. 48, no. 12, pp. 3395–3406, 2000.

[39] Ong, S.H., "Signal-adapted wavelets for Doppler radar system," in *7th International Conference on Control, Automation, Robotics and Vision, 2002. ICARCV 2002.*, vol. 1. Nanyang Technological Univ, 2002, pp. 19–23. [Online]. Available: http://ieeexplore.ieee.org/document/1234783/

[40] R. X. Gao and R. Yan, *Wavelets: Theory and applications for manufacturing*, 1st ed. Boston, MA: Springer US, 2011. [Online]. Available: http://link.springer.com/10.1007/978-1-4419-1545-0

[41] R. C. Guido, J. F. W. Slaets, R. Köberle, L. O. B. Almeida, and J. C. Pereira, "A new technique to construct a wavelet transform matching a specified signal with applications to digital, real time, spike, and overlap pattern recognition," *Digital Signal Processing: A Review Journal*, vol. 16, no. 1, pp. 24–44, 2006.

[42] S. Maslakovic, I. R. Linscott, M. Oslick, and J. D. Twicken, "Smooth orthonormal wavelet libraries: Design and application," *ICASSP, IEEE International Conference on Acoustics, Speech and Signal Processing - Proceedings*, vol. 3, pp. 1793–1796, 1998.

[43] S. Maslakovic, I. Linscott, M. Oslick, and J. Twicken, "Excising radio frequency interference using the discrete wavelet transform," in

Proceedings of Third International Symposium on Time-Frequency and Time-Scale Analysis (TFTS-96), vol. 2. IEEE, 1996, pp. 349–352. [Online]. Available: http://citeseerx.ist.psu.edu/viewdoc/summary?doi=10.1.1.53.1869http://ieeexplore.ieee.org/document/547485/

[44] P. Desarte, B. Macq, and D. Slock, "Signal-adapted multiresolution transform for image coding," *IEEE Transactions on Information Theory*, vol. 38, no. 2, pp. 897–904, mar 1992. [Online]. Available: http://ieeexplore.ieee.org/document/119749/

[45] P. Moulin, "A new look at signal-adapted QMF bank design," in *1995 International Conference on Acoustics, Speech, and Signal Processing*, vol. 2. IEEE, 1992, pp. 1312–1315. [Online]. Available: http://ieeexplore.ieee.org/document/480481/

[46] A. H. Tewfik, D. Sinha, and P. Jorgensen, "On the Optimal Choice of a Wavelet for Signal Representation," *IEEE Transactions on Information Theory*, vol. 38, no. 2, pp. 747–765, 1992.

[47] S. A. Mallat, "Time-Frequency Dictionaries," *A Wavelet Tour of Signal Processing*, pp. 11–14, 1999.

[48] J. P. Li and Y. Y. Tang, "General Analytic Construction for Wavelet Low-Passed Filters," 2001, pp. 314–320. [Online]. Available: http://link.springer.com/10.1007/3-540-45333-4{_}38

[49] J. P. Li, Y. Y. Tang, Z. H. Yan, and W. P. Zhang, "Uniform Analytic Construction of Wavelet Analysis Filters Based on Sinc and Cosine Trigonometric Functions," pp. 569–585, 2001.

[50] J. Odegard and C. Burrus, "Discrete finite variation: A new measure of smoothness for the design of wavelet basis," *Acoustics, Speech, and Signal*, pp. 1467–1470, 1996. [Online]. Available: http://ieeexplore.ieee.org/xpls/abs{_}all.jsp?arnumber=543939

[51] H. Zou and A. H. Tewfik, "Parametrization of Compactly Supported Orthonormal Wavelets," *IEEE Transactions on Signal Processing*, vol. 41, no. 3, pp. 1428–1431, 1993.

[52] C. J. Willmott and K. Matsuura, "On the use of dimensioned measures of error to evaluate the performance of spatial interpolators," *International Journal of Geographical Information Science*, vol. 20, no. 1, pp. 89–102, 2006.

[53] A. R. Thompson, J. M. Moran, and G. W. Swenson, *Interferometry and Synthesis in Radio Astronomy*, 2017. [Online]. Available: http://link.springer.com/10.1007/978-3-319-44431-4

[54] H. Abdi and P. Molin, "Lilliefors / Van Soest's test of normality," Univeristy of Texas in Dallas, Dallas Texas, Tech. Rep., 1967. [Online]. Available: https://www.utdallas.edu/{~}herve/Abdi-Lillie2007-pretty.pdf

[55] J. Berger, R. R. Coifman, and M. J. Goldberg, "Removing Noise from Music Using Local Trigonometric Bases and Wavelet Packets," *J. Audio Eng. Soc*, vol. 42, no. 10, pp. 808–818, 1994. [Online]. Available: http://www.aes.org/e-lib/browse.cfm?elib=6926

[56] A. Wilkinson, *Notes on Radar/Sonar Signal Processing: Fundamentals.* Cape Town: University of Cape Town, 2016, vol. 2016.

[57] SKA, "MeerKAT," 2013. [Online]. Available: http://www.ska.ac.za/gallery/kat-7/

[58] J. D. Kraus, *Radio Astronomy*, 2nd ed. New York, NY, USA: McGraw-Hill, 1966. [Online]. Available: papers://b340de6f-f85e-427d-b935-938457f3fdc7/Paper/p4263

[59] N. Oozeer, "Polarisation in radio astronomy," Square Kilometer Array, Cape Town, Tech. Rep., 2018.

[60] H. J. van Langevelde, "Interferometry concepts," JIVE, Leiden, Tech. Rep., 2015. [Online]. Available: http://www.eas-journal.org/10.1051/eas/1569003

[61] N. Oozeer, M. Bietenholz, and S. Goedhart, "Introduction To Casa:A KAT-7 Data Reduction Guide ."

[62] J. P. McMullin, B. Waters, D. Schiebel, W. Young, and K. Golap, "CASA Architecture and Applications," *Astronomical Society of the Pacific Conference Series*, vol. 376, p. 127, 2007.

[63] E. Abebe, "A Study Of Potential Calibrators Using The KAT-7 Radio Telescope," Ph.D. book, Univesity of Cape Town, 2002.

[64] E. Lakdawalla, "Pretty picture: Messier 83," 2010. [Online]. Available: http://www.planetary.org/blogs/emily-lakdawalla/2010/2499.html

[65] ICASA, "Inmarsat comments on the ICASA consultation," Tech. Rep., 2012. [Online]. Available: http://www.erodocdb.dk/doks/doccategoryECC.aspx?doccatid=1

[66] N. Mohan and D. Rafferty, "PyBDSF: Python Blob Detection and Source Finder," 2015. [Online]. Available: http://adsabs.harvard.edu/abs/2015ascl.soft02007M

[67] SKA, "About MeerKAT – SKA SA." [Online]. Available: https://www.ska.ac.za/science-engineering/meerkat/about-meerkat/

Appendix A

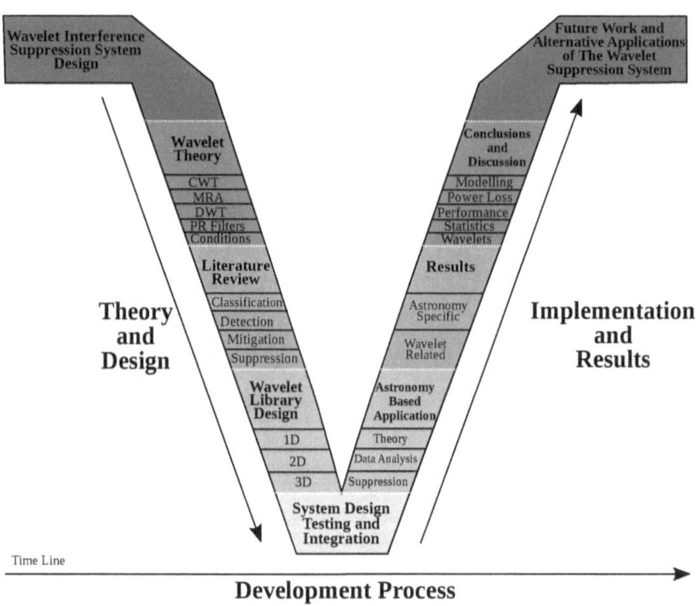

Figure A.1: V Diagram Showing the Methodology of the Research Project

Appendix B

TABLE SHOWING THE CRITICAL VALUES OF THE LILLEFORS TEST WITH $K = 100000$ SAMPLES. THE LAST ROW SHOWS THE CRITICAL VALUE OF EACH TEST GIVEN SOME CONFIDENCE INTERVAL, WHERE $f_N = \dfrac{0.83 + N}{\sqrt{N}} - 0.1$ [54]

N	$\alpha = 0.2$	$\alpha = 0.15$	$\alpha = 0.1$	$\alpha = 0.05$	$\alpha = 0.01$
4	0.3027	0.3216	0.3456	0.3754	0.4129
5	0.2893	0.3027	0.3188	0.3427	0.3959
6	0.2694	0.2816	0.2982	0.3245	0.3728
7	0.2521	0.2641	0.2802	0.3041	0.3504
8	0.2387	0.2502	0.2649	0.2875	0.3331
9	0.2273	0.2382	0.2522	0.2744	0.3162
10	0.2171	0.2273	0.2410	0.2616	0.3037
11	0.2080	0.2179	0.2306	0.2506	0.2905
12	0.2004	0.2101	0.2228	0.2426	0.2812
13	0.1932	0.2025	0.2147	0.2337	0.2714
14	0.1869	0.1959	0.2077	0.2257	0.2627
15	0.1811	0.1899	0.2016	0.2196	0.2545
16	0.1758	0.1843	0.1956	0.2128	0.2477
17	0.1711	0.1794	0.1902	0.2071	0.2408
18	0.1666	0.1747	0.1852	0.2018	0.2345
19	0.1624	0.1700	0.1803	0.1965	0.2285
20	0.1589	0.1666	0.1764	0.1920	0.2226
21	0.1553	0.1629	0.1726	0.1881	0.2190
22	0.1517	0.1592	0.1690	0.1840	0.2141
23	0.1484	0.1555	0.1650	0.1798	0.2090

Continued on next page

24	0.1458	0.1527	0.1619	0.1766	0.2053
25	0.1429	0.1498	0.1589	0.1726	0.2010
26	0.1406	0.1472	0.1562	0.1699	0.1985
27	0.1381	0.1448	0.1533	0.1665	0.1941
28	0.1358	0.1423	0.1509	0.1641	0.1911
29	0.1334	0.1398	0.1483	0.1614	0.1886
30	0.1315	0.1378	0.1460	0.1590	0.1848
31	0.1291	0.1353	0.1432	0.1559	0.1820
32	0.1274	0.1336	0.1415	0.1542	0.1798
33	0.1254	0.1314	0.1392	0.1518	0.1770
34	0.1236	0.1295	0.1373	0.1497	0.1747
35	0.1220	0.1278	0.1356	0.1478	0.1720
36	0.1203	0.1260	0.1336	0.1454	0.1695
37	0.1188	0.1245	0.1320	0.1436	0.1677
38	0.1174	0.1230	0.1303	0.1421	0.1653
39	0.1159	0.1214	0.1288	0.1402	0.1634
40	0.1147	0.1204	0.1275	0.1386	0.1616
41	0.1131	0.1186	0.1258	0.1373	0.1599
42	0.1119	0.1172	0.1244	0.1353	0.1573
43	0.1106	0.1159	0.1228	0.1339	0.1556
44	0.1095	0.1148	0.1216	0.1322	0.1542
45	0.1083	0.1134	0.1204	0.1309	0.1525
46	0.1071	0.1123	0.1189	0.1293	0.1512
47	0.1062	0.1113	0.1180	0.1282	0.1499
48	0.1047	0.1098	0.1165	0.1269	0.1476
49	0.1040	0.1089	0.1153	0.1256	0.1463
50	0.1030	0.1079	0.1142	0.1246	0.1457
< 50	$\dfrac{0.741}{f_N}$	$\dfrac{0.755}{f_N}$	$\dfrac{0.819}{f_N}$	$\dfrac{0.895}{f_N}$	$\dfrac{1.035}{f_N}$

Appendix C

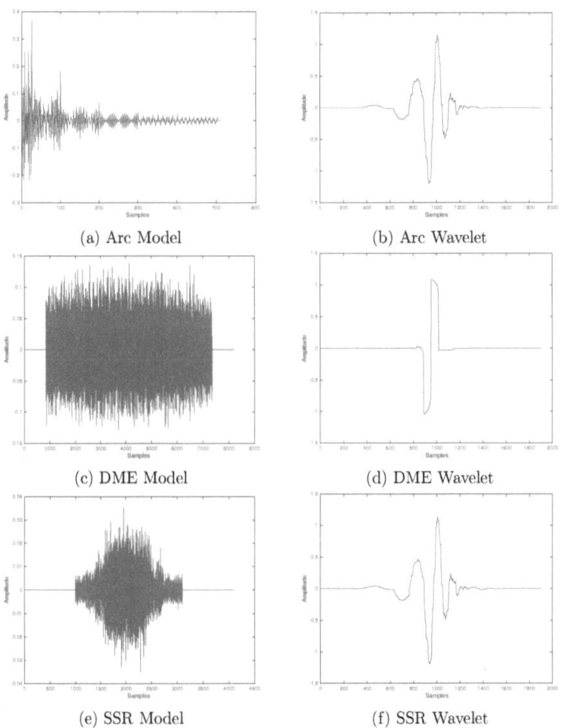

(a) Arc Model

(b) Arc Wavelet

(c) DME Model

(d) DME Wavelet

(e) SSR Model

(f) SSR Wavelet

Figure C.1: Figures Showing the Selected Wavelet for Each Median Modeled Signal

Appendix D

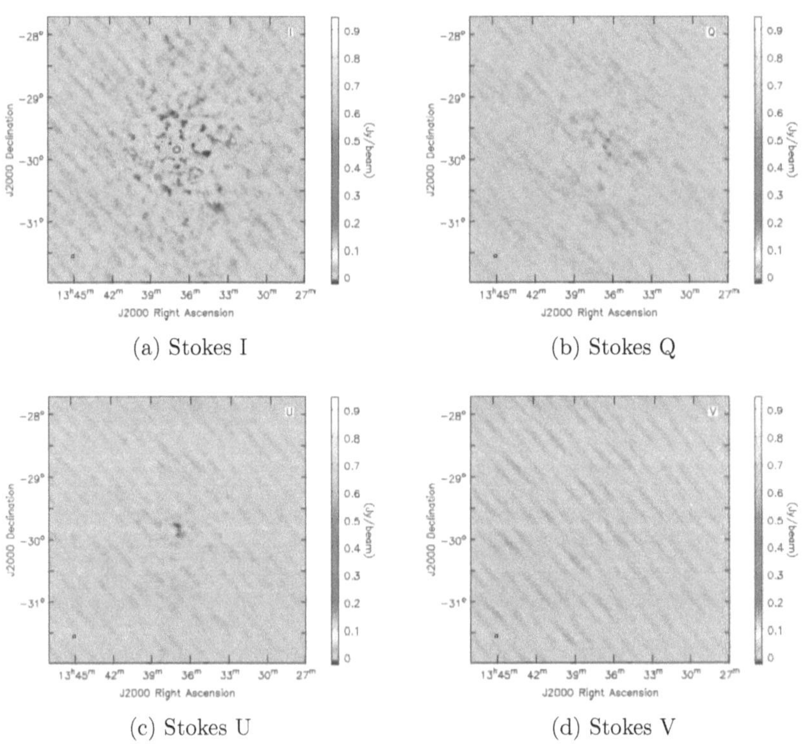

(a) Stokes I

(b) Stokes Q

(c) Stokes U

(d) Stokes V

Figure D.1: Figures the Messier 83 Galaxy at Each Polarization

Appendix E

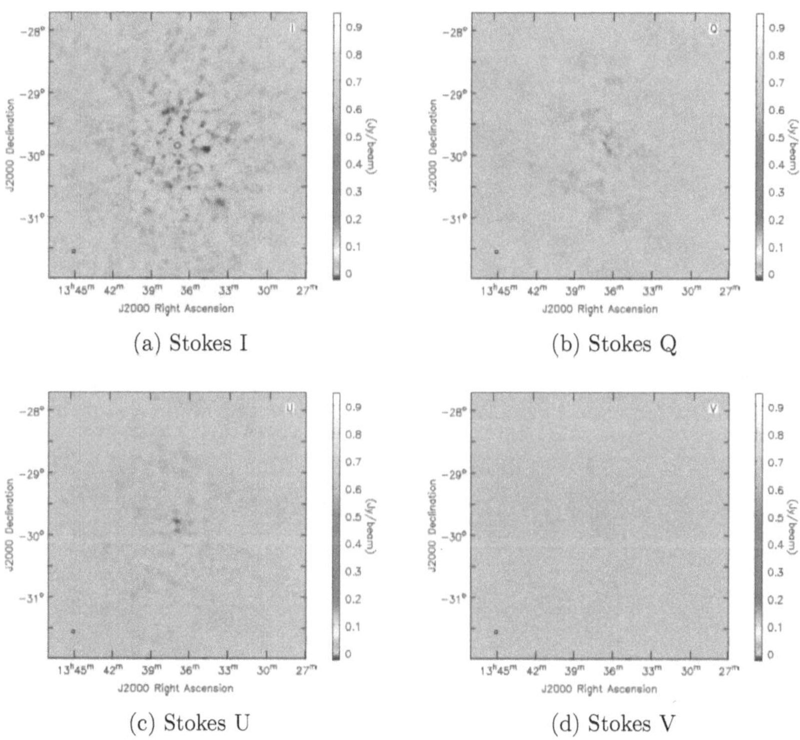

(a) Stokes I

(b) Stokes Q

(c) Stokes U

(d) Stokes V

Figure E.1: Figures showing the Flagged Images of Messier 83 Galaxy at Each Polarization

Appendix F

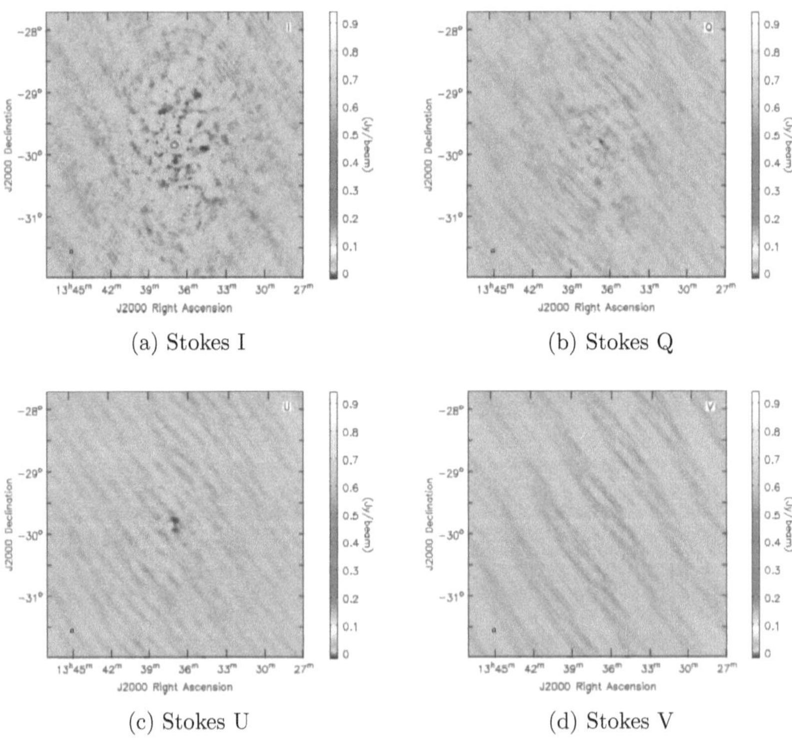

(a) Stokes I

(b) Stokes Q

(c) Stokes U

(d) Stokes V

Figure F.1: Figures of RFI Suppressed Messier 83 Galaxy at Each Polarization

APPLICATION FORM

Please Note:

Any person planning to undertake research in the Faculty of Engineering and the Built Environment (EBE) at the University of Cape Town is required to complete this form **before** collecting or analysing data. The objective of submitting this application *prior* to embarking on research is to ensure that the highest ethical standards in research, conducted under the auspices of the EBE Faculty, are met. Please ensure that you have read, and understood the **EBE Ethics in Research Handbook** (available from the UCT EBE, Research Ethics website) prior to completing this application form: http://www.ebe.uct.ac.za/ebe/research/ethics1

APPLICANT'S DETAILS

Name of principal researcher, student or external applicant		Michael Mesarcik
Department		Electrical Engineering
Preferred email address of applicant		msrmic004@myuct.ac.za
If Student	Your Degree: e.g., MSc, PhD, etc.	MSc
	Credit Value of Research: e.g., 60/120/180/360 etc.	120
	Name of Supervisor (if supervised):	Associate Professor Daniel O'Hagan
If this is a researchcontract, indicate the source of funding/sponsorship		
Project Title		A Real-Time Wavelet based Noise Suppression System on a FPGA

I hereby undertake to carry out my research in such a way that:

- there is no apparent legal objection to the nature or the method of research; and
- the research will not compromise staff or students or the other responsibilities of the University;
- the stated objective will be achieved, and the findings will have a high degree of validity;
- limitations and alternative interpretations will be considered;
- the findings could be subject to peer review and publicly available; and
- I will comply with the conventions of copyright and avoid any practice that would constitute plagiarism.

SIGNED BY	Full name	Signature	Date
Principal Researcher/ Student/External applicant	Michael Mesarcik		15/11/2017

APPLICATION APPROVED BY	Full name	Signature	Date
Supervisor (where applicable)	Willem Petrus Francis Schonken *(handwritten)* Click here to enter text.	*(signature)*	5 Feb 20.. *(handwritten)* Click here to enter a date.
HOD (or delegated nominee) Final authority for all applicants who have answered NO to all questions in Section1; and for all Undergraduateresearch (Including Honours).	Olabisi Falowo. *(handwritten)* Click here to enter text.	*(signature)*	20/10/2018 *(handwritten)* Click here to enter a date.
Chair : Faculty EIR Committee For applicants other than undergraduate students who have answered YES to any of the above questions.			